Handbook of QS-9000 Tooling and Equipment Certification

Other SAE books of interest:

QS-9000 Quality Systems Handbook
by David Hoyle
(Order No. R-195)

For more information or to order this book, contact SAE at 400 Commonwealth Drive, Warrendale, PA 15096-0001; (724) 776-4970; fax (724) 776-0790; e-mail: publications@sae.org; http://www.sae.org/BOOKSTORE.

Handbook of QS-9000 Tooling and Equipment Certification

George Mouradian

Society of Automotive Engineers, Inc.
Warrendale, Pa.

•

Newnes
An imprint of Butterworth-Heinemann

| Oxford | London | Boston | Munich | New Delhi |
| Singapore | Sydney | Tokyo | Toronto | Wellington |

Copyright © 2000   Society of Automotive Engineers, Inc.
400 Commonwealth Drive, Warrendale, PA 15096-0001 U.S.A.
Phone: (724)776-4841; Fax: (724)776-5760
E-mail: publications@sae.org; http://www.sae.org

Library of Congress Cataloging-in-Publication Data

Mouradian, George
  Handbook of QS-9000 tooling and equipment certification/George
Mouradian.
    p.cm.
  Includes bibliographical references and index.
  ISBN 0-7680-0527-2
  1. Industrial equipment--Certification. 2. Machine-tools--Design and
construction. 3. Machine-tools--Reliability. 4. QS-9000 (Standard)
  I. Title.

  TS191 .M68 2000
  629.23'4'0685--dc21   99-089646

ISBN 0-7680-0527-2

SAE Order No. R-265

Butterworth-Heinemann Ltd.
Linacre House, Jordan Hill, Oxford OX2 8DP

A member of the Reed Elsevier plc group

British Library Cataloguing in Publication Data
A catalogue record for this book is available from the British Library

Butterworth-Heinemann ISBN 0 7506 4773 6

Table of Contents

Foreword

The intent of this book is to provide manufacturers of tooling and equipment some guidance in attaining certification of the reliability and maintainability (R&M) requirements of *QS-9000 Tooling and Equipment (TE) Supplement*. The book also will be helpful to users of tooling and equipment in understanding the requirements of the *TE Supplement*. Since 1987 when the European companies initiated the requirement that their suppliers had to become ISO 9000 certified, the certification process has mushroomed into what it is today. I think the process will continue to grow.

Although European organizations started the process, it definitely has expanded to the international arena. Offshoots to the ISO 9000 standards such as QS-9000, the *QS-9000 Tooling and Equipment Supplement*, AS9000, TL9000, ISO 14000, and others yet to be developed and required are becoming standards for numerous industries. This handbook covers one aspect of the certification process that includes many additional requirements above and beyond the QS-9000 implementation. It presents an overall picture but also provides the methods and techniques necessary to obtain the TE registration.

Chapter 1 elaborates on the history behind ISO 9000 and on the standards that followed it. The chapter then briefly provides an explanation of the *TE Supplement* requirements, stating what must be accomplished to obtain the certification. Chapter 1 reviews each of the elements and Customer-specific Requirements of the *TE Supplement* and then comments on their provisions.

The chapters that follow explain the techniques and skills that are necessary for tooling and equipment manufacturers to become certified. As expected, most of what is presented focuses on the reliability and maintainability (R&M) routines. Some of the techniques are more complex and difficult than others, but most of what is necessary is presented in fine detail, so that an organization will be able to utilize the handbook to its fullest advantage.

Another purpose of this handbook is to indoctrinate manufacturers and users to work in harmony. The book also can be used as a training manual. Both parties should understand the innards of R&M and how they should be utilized. The ultimate goal of the manufacturer is to provide machinery to the user that meets all the specifications and requirements, and is cost effective.

George Mouradian
December 3, 1999

Chapter 1

QS-9000 Tooling and Equipment Supplement Background and Certification

In 1993, the Society of Automotive Engineers, Inc. (SAE) and the National Center for Manufacturing Sciences, Inc. (NCMS) published the *Reliability and Maintainability Guideline for Manufacturing Machinery and Equipment.* The guideline consolidated the reliability and maintainability (R&M) terminology, methodology, and procurement language for machine suppliers and users of tooling and equipment (TE) appropriated for the manufacture of unique components. The guideline outlined the requirements for R&M to enable suppliers and users of TE to understand the basics of the required R&M skills. One of the intents of the guideline is to encourage the partnership between the supplier and user, ensuring they both comprehend that all parties should be involved in the continuous improvement of equipment operation and design. The successful implementation of an R&M program requires commitment to the process by both the supplier and the user.

Before the *R&M Guideline* was published, Ford Motor Company had a company guideline that served as an initial pathfinder for the R&M skills and techniques. The company guideline was to be implemented in several Ford operating units. Later, General Motors and Chrysler developed their own guidelines, to be used by their own units and also among various parts manufacturers. Many of the techniques related directly to the application of the reliability engineering practices to machine building.

The SAE and NCMS acknowledge the pioneering efforts of the automotive industry and three organizations—NCMS, The Industrial Technical Institute (ITI), and Western Michigan University—in their efforts to provide an R&M guidance document that could be used by any organization interested in learning and using the R&M skills and disciplines. Other companies provided key personnel who collectively spent many hours studying the complexities of the practice of R&M and ensuring that the wording, structure, and organization would be understood by interested parties. It was extremely important that the manufacturing companies comprehended the forum and environment necessary for their organizations and for the end users.

This handbook is designed to assist TE manufacturers in meeting the R&M quality system requirements of the second edition of *Quality System Requirements— Tooling and Equipment Supplement* (also referred to here as QS-9000TE or *TE Supplement*). In 1996, Chrysler Corporation, Ford Motor Company, and General Motors Corporation (the Big Three) published the first edition of *Quality System Requirements—Tooling and Equipment Supplement*, which provided an interpretation of the QS-9000TE requirements. Riveria Tool & Die assisted in the development of the supplement. This handbook provides the basic resources and advice necessary for TE businesspersons to achieve the QS-9000TE registration. The handbook can be used as a support text for the Big Three *TE Supplement.*

In effect, the *TE Supplement* commonizes the quality system requirements for organizations wishing to acquire the TE certification. The Big Three have agreed that the *TE Supplement* replaces Chrysler's TESQA (Tooling and Equipment Supplier Quality Assurance) and Ford's Facilities & Tools Quality Systems Survey (F&T QSS). General Motors, which had its own guideline, accepted the unification of a common guideline, believing it would greatly benefit expansion of the practice of R&M among the manufacturing community.

Organizations wishing to become certified to the *TE Supplement* must implement the requirements of QS-9000 in their entirety, but with the exception of the ISO 9000 applicable elements. The *TE Supplement* specified requirements are in addition to QS-9000, and suppliers wishing to obtain the TE certification must comply with them. Both suppliers and users should understand the specified requirements.

For those who are relatively new to the ISO scenario, a little history of what has been happening could be helpful. In the early 1980s, some European nations had separate quality standards that they recommended be used by the companies within their borders. Some countries enforced the standards by legal means. In the United States, the Department of Defense has several military standards that it imposed on its contractors. Good examples of government standards that were required contractual documents are MIL-Q-9858, MIL-I-45208, MIL-STD-470, and MIL-STD-785. Other organizations had their own standards that they compelled their suppliers to use.

In the mid-1980s, European countries came together and decided on some unity of purpose. There is a long story on the development of the European Economic Union and on some of the other unification bodies. This handbook will address only the quality standards aspect of the numerous committees that met together for years before a unified standard was issued. When the International Organization for Standardization (ISO) first released the ISO 9000 series standards, many companies perceived the nations that initiated the documents were European. In some manner, the series standards originally were European, but they rapidly acquired an international flavor. In 1987, the ISO 9000 quality standard series was published officially for the international community. Before the final release of the series, several years of preliminary draft reviews were done by the member nations. The National Institute of Standards and Technology (NIST) represented the United States in this effort.

Five quality documents were initially released in 1987 by the ISO. Copies also can be obtained from the American Society for Quality (ASQ, formerly American Society for Quality Control [ASQC], Milwalukee, Wisconsin). In 1994, the documents all were revised, improved, and released again as follows:

1. ISO 9000 (ANSI/ISO/ASQC Q9000-1994 Series)
   Quality Management and Quality Assurance Standards: Guidelines for the Selection and Use

2. ISO 9001 (ANSI/ISO/ASQC Q9001-1994)
   Quality Systems: Model for Quality Assurance in Design, Development, Production, Installation, and Servicing

3. ISO 9002 (ANSI/ISO/ASQC Q9002-1994)
    Quality Systems: Model for Quality Assurance in Production, Installation, and Servicing

4. ISO 9003 (ANSI/ISO/ASQC Q9003-1994)
    Quality Systems: Model for Quality Assurance in Final Inspection and Test

5. ISO 9004 (ANSI/ISO/ASQC Q9004-1994)
    Quality Management and Quality Systems Elements: Guidelines

Before the second release, the ISO circulated to all the member nations the draft revisions for comment and improvement. Several changes were made to the documents. The ISO numbering system remained the same. In the United States, the ANSI/ASQ numbers were changed to the ISO system (e.g., ANSI/ASQC Q91 was revised to Q9001). Some member nations put their own designations on their standards, but all the standards basically were equivalent to the international standards. As of this writing, there are more than 100 ISO member nations. A few of the more basic revisions to the 1994 documents included the following:

*   More focus on the needs of the customer

*   Numbering system and sequence of clauses now are consistent in the three auditable standards (e.g., 9001, 9002, and 9003)

*   Subtitles more accurately reflect the text of the subclauses

*   Text is easier to understand

*   A quality manual and quality planning now are required

*   The scope of the standards is expanded to recognize more common usage

*   Documented design reviews now are required and are to include representatives of all functions concerned

Several additional changes were made to the 1994 version, but those described here were the most basic ones.

As of the publication of *Handbook of QS-9000 Tooling and Equipment Certification*, the ISO9000 documents are currently under revision. Release is planned for sometime in the year 2000.

In 1994, the Big Three and several trucking companies also accepted the international standards, but they felt the requirements were inadequate for their needs. The Big Three and the truckers thought the ISO quality standards were too generic for their purposes. Thus, the standard *Quality System Requirements QS-9000* was published. Essentially, the QS-9000 adopted all the requirements of the ISO standards and added specific automotive addendum requirements to each of the 20 ISO elements plus Section II: Sector-specific Requirements and Section III: Customer-specific Requirements. Chrysler required all its first-tier suppliers to be certified to the applicable QS-9000 requirements by July 31, 1997. General Motors put the deadline for its suppliers at December 31, 1997. As of this writing, Ford does not require certification of its current suppliers; however, Ford does require compliance (i.e., suppliers must demonstrate that their quality systems are in place with regard to the QS-9000 requirements). In addition, Ford requires new suppliers and those on probation to be registered.

In July 1997, a new document entered the picture, *Quality System Requirements— Tooling and Equipment Supplement*. In June 1998, the second edition was published. As noted previously, the document was intended primarily for suppliers of tooling and equipment. This handbook concentrates on the second edition R&M requirements of the *TE Supplement*. A review of the R&M requirement elements and the added Section II: Customer-specific Requirements is presented here. The first edition has Section I: ISO 9000-Based Requirements (20 elements), plus Section II: Sector-specific Requirements, and Section III: Customer-specific Requirements. The second edition incorporated the Sector-specific Requirements into each of their appropriate elements in Section I. Each of the 20 plus 3 QS-9000 elements of the *Quality System Requirements—Tooling and Equipment Supplement* depicts the abbreviated details of each element. In the following summary, the element number is followed by its description.

4.1   Management Responsibility

4.1.1   Quality Policy*

Quality objectives shall include:
- Reliability
- Maintainability
- Durability

4.2   Quality Planning

4.2.3   Quality Planning

4.2.3.1   Advanced Product Quality Planning
- The TE suppliers shall advance quality planning process embracing R&M through Life Cycle Costing (LCC).
- The supplier shall fully understand the quality/reliability/maintainability/durability requirements of the customer. The intention is to accomplish reliability and maintainability methodologies through multidisciplinary function exercise.

4.2.3.2   Special Characteristics
Controls shall be devised and implemented throughout all of the development phases. Controls shall be documented in a control plan.

4.2.3.2   Process Failure Modes and Effects Analysis (Process FMEAs)
- Use of FMEAs shall be documented on all manufactured products.
- The *FMEA Manual* and the *R&M Guideline* should be used for guidance.
- Internal processes should be addressed using the FMEA discipline.
- FMEAs shall be living documents and shall be reviewed and updated when changes in the product or process occur.

---

* Note: Throughout this supplement, references to quality shall include reliability, maintainability, and durability.

4.2.3.7   The Control Plan
- Suppliers shall prepare a control plan using a multidisciplinary approach. The Advanced Product Quality Planning and Control Plan provides an example.
- The output of the advanced quality planning process, beyond the development of robust processes, is a Control Plan.
- The Control Plan shall be a living document and shall be reviewed and updated when changes occur.
- Customer approval of the Control Plan may be required.

4.2.4   Product Approval Process (PPAP), Machinery Qualification Requirements (The Product Part Approval Process in QS-9000 does not apply to TE suppliers. The following section, Machinery Qualification Runoff Requirements, replaces the PPAP.)

4.2.4.1   Purpose
There must be assurance that equipment purchased by the customer will be of acceptable quality when received. All identifiable problems shall be eliminated prior to the machinery being integrated into a larger system or installed at the customer's facility. The intent of this procedure is to:
- Reduce or eliminate startup delays
- Improve the quality of all components and systems to a level that conforms to customer standards
- Resolve software and control problems prior to launch
- Confirm that equipment cycle time will meet the customer's productivity requirements
- Verify reliability of tooling and equipment

4.2.4.2   Machinery Qualification Runoff Requirements
Any failure during Machinery Qualification Runoff shall be documented with root cause analysis by the supplier.

4.2.4.3   Procedure

4.2.4.3.a   50/20 Dry Run
The 50-hour quality test has been established for robots. At the customer's discretion, the OEM's test data may be used in lieu of an on-site 50-hour quality test. The 20-hour continuous dry run applies to all machinery, including all robots within these systems. The following steps of this procedure govern both the 50-hour and 20-hour applications.

- All tests shall be conducted at the facility of the system supplier.
- Customer personnel will be on site at the beginning of each test and may provide assistance.
- The supplier shall provide the resources to run and service the test.
- The supplier shall be responsible for all components during the test runs including scheduled maintenance.
- Time required to comply with this program shall be recognized as part of the schedule for delivery.
- Failure during the 20-hour run shall require a "re-start" of the 20-hour requirement, unless waived by the customer.

Tooling may be exempt from the 20-hour continuous dry run requirement.

4.2.4.3.b   Phase 1 of Initial Process Performance—Preliminary Evaluation
A sample run will be the first formalized process evaluation with approved parts. The customer does not have to be present; however, all parts should be identified in sequence and should be available for review by customer personnel.

4.2.4.3.c   Phase 2 of Initial Process Performance—Pp
Pp is a measure of process irrespective of tolerance location. This evaluation will be conducted at the supplier's location, possibly with a customer representative. The supplier shall consult with the customer representative for clarification of:
- Sample size and type
- Frequency
- Total quantity
- Customer representation requirements
- Acceptance criteria

On processes with inherently low variation, target nominals may be purposely shifted in consideration for tool wear.

4.2.4.3.d   Phase 3 of Initial Process Performance—Ppk
Ppk is a measure of process performance and how it relates to tolerance location. This evaluation will be conducted at the supplier's location, possibly with a customer representative. If the Ppk from the Phase 2 data is acceptable, then the Phase 3 Ppk requirement has been simultaneously met. The supplier shall consult with the customer representative for clarification of:
- Sample size and type
- Frequency
- Total quantity
- Customer representation requirements
- Acceptance criteria

On processes with inherently low variation, target nominals may be purposely shifted.

4.2.4.3.e   Reliability
The final part of the pre-delivery acceptance procedure is a projection of machinery reliability by the supplier who shall demonstrate that the machinery is designed and built to the reliability specification.

4.2.4.3.f   20-Hour Dry Run
To verify the as-installed condition of the machinery, the 20-hour dry run is repeated with the cycle speeds continuously for the 20 hours. Both customer and supplier should be present.

4.2.4.3.g   Short-Term Process Study
This check is to be performed under production conditions. Data from the 25 subgroups (recommend 125 pieces) are to be collected and recorded. Specific data summarization may be specified by the customer.

4.2.4.3.h   Long-Term Process Study
The long-term process study is conducted at the customer's plant per the requirements of the individual plant.

4.2.5.3   Techniques for Continuous Improvement
The supplier also shall demonstrate knowledge of the following measures:
- Mean Time Between Failures (MTBF)
- Mean Time To Repair (MTTR)
- Life Cycle Cost (LCC)
- Reliability Growth

Cross-Functional Teams
- Implementation of the R&M methodologies is intended to be a cross-functional exercise.

Feasibility Reviews
- Feasibility reviews shall be documented.

4.3   Contract Review
No R&M requirements were added to Element 4.3.

4.4   Design Control

    4.4.2   Design and Development Planning

        4.4.2.1   Required Skills
        The required skills of Mean Time To Repair (MTTR), Mean Time Between Failures (MTBF), Fault Tree Analysis (FTA), Life Cycle Cost (LCC), and environmental characterization should be utilized as required and understood.

    4.4.5   Design Output

        4.4.5.1   Design Output—Supplemental
        Design output also shall include analysis of test data and projections of reliability, maintainability, durability, and Life Cycle Cost (LCC).

    4.4.9   Design Changes

        4.4.9.1   Design Changes—Supplemental
        Suppliers shall maintain a log of all design changes throughout each phase of the machinery build, including those requested by the customer.

    4.4.10   Customer Prototype Support
- Performance tests shall consider and include, as appropriate, R&M, durability, and product life cycle.
- Suppliers shall engage a comprehensive prototype program and/or utilize predictive R&M techniques.
- Accelerated life tests shall be conducted on crucial components.

4.5   Document and Data Control
No R&M requirements were added to Element 4.5.

4.6　Purchasing

    4.6.2　Evaluation of Subcontractors

        4.6.2.1　Subcontractor Development
        Subcontractor development shall use the *TE Supplement* Sections I and II as the fundamental quality system requirement.

4.7　Control of Customer-Supplier Product
No R&M requirements were added to Element 4.7.

4.8　Product Identification and Traceability
The supplier shall establish and maintain a tracking system for components and sub-assemblies. An effective system shall identify components to their next operation and by their job number. The tracking system should also cross-reference engineering drawings and bill of materials or equivalent.

4.9　Process Control

    4.9.1　Process Monitoring and Operator Instructions
    The use of process monitoring and operator instructions shall be adequate to document the TE supplier's process. Process monitoring and instructions may take the form of several records. Each employee shall be familiar with the work instructions and the objective of the job assignment.

    4.9.2　Maintaining Process Control
    This section replaces the PPAP found in QS-9000 with the Machinery Qualification Runoff Requirements, 4.3.4.

    4.9.7　Appearance Items
    This section is available only for reference.

4.10 Inspection and Testing

    4.10.4 Final Inspection and Testing

        4.10.4.1 Layout Inspection and Functional Testing
               Functional verification (to applicable customer standards) shall be performed for the TE Requirement. The Qualification Runoff Requirements table on page 20 of the *TE Supplement* summarized the testing required for the final buyoff. The results shall be available for the customer's review.

4.11 Control of Inspection, Measuring, and Test Equipment

    4.11.4 Measuring System Analysis
        Note added: All hand-held variable gages that do not require a setup are exempt from this requirement (e.g., micrometers, calipers, and height gages). These gages still require gage identification, calibration, and traceability.

4.12 Inspection and Test Status
    No R&M requirements were added to Element 4.12.

4.13 Control of Nonconforming Product

    4.13.3 Control of Reworked Product
        No R&M requirements were added; however, this section is available for reference.

    4.13.4 Engineering Approved Product Authorization
        No R&M requirements were added; however, this section is available for reference.

4.14 Corrective and Preventive Action
    No R&M requirements were added to Element 4.14.

4.15 Handling, Storage, Packing, Preservation, and Delivery

4.15.6 Delivery

4.15.6.1 Supplier Delivery Performance Monitoring
- The supplier shall establish a goal of 100% on-time shipments to meet the customer's requirements. The supplier shall develop timing plans which can be used for management planning and control and can be adjusted to show the effect of changes when needed. Critical path scheduling shall be required or timing of complex manufacturing systems.
- Scheduling systems shall be in place, which accurately control timing for the manufacturing of all major components and assemblies, for equipment test, runoff, installation, and tryout. The scheduling systems shall include subcontractors and shall correspond with material procurement timing and capacity resource planning. Capacity limitations for machines, labor, and facilities should be identified to allow alternate plans when needed.

- Regular timing meetings should be held to determine the status of actions needed to maintain timing and effect of engineering changes.
- Note: Gantt charts are suggested for providing good representation of the overall plan.

4.15.6.2 Production Scheduling
No R&M requirements were added; however, this section is available for reference.

4.15.6.3 Electronic Communication
No R&M requirements were added; however, this section is available for reference.

4.15.6.4 Shipment Notification System
No R&M requirements were added; however, this section is available for reference.

4.16 Control of Quality Records
No R&M requirements were added to Element 4.16.

4.17 Internal Audits
No R&M requirements were added to Element 4.17.

4.18 Training
- The supplier shall implement a formal training program to include R&M.
- Note: The *Reliability and Maintainability Guideline* provides an example of a formal training program.

4.19 Servicing

4.19.1 Feedback of Information from Service
A procedure for communication of information on machine uptime, R&M history, and service concerns to manufacturing, as well as design activities, shall be established and maintained.

4.20 Statistical Techniques

4.20.1 Identification of Need
Statistical techniques also shall include reliability and maintainability analysis.

4.20.2 Procedures: Knowledge of Statistical Concepts
Statistical concepts shall include:
- Mean Time Between Failures (MTBF)
- Mean Time To Repair (MTTR)
- Short-run Statistical Process Control (SPC)
- Control charts*
- p, np, c, and u charts or any other appropriate statistical techniques*

---

* Reference the Statistical Process Control (SPC) manual.

Section II: Customer-specific Requirements

This section is subdivided into specific requirements for Chrysler, Ford, General Motors, (Big Three) and other OEMs. Requirements for each organization are as follows.

Chrysler-specific Requirements

The following sections of QS-9000 do not apply to tooling and equipment suppliers and are only for reference.

- Product Creation Process
- Annual Layout
- Product Verification/Design Verification
- Appearance Masters
- Packaging, Shipping, and Labeling
- Process Approval

Ford-specific Requirements

The following sections of QS-9000 do not apply to tooling and equipment suppliers and are only for reference.

- Third Party Registration Requirements
- Control Item (∇) Parts
- Annual Layout
- Setup Verification
- Control Item (∇) Fasteners
- Heat Treating
- Process Changes and Design Changes for Supplier-Responsible Designs
- Supplier Notification of Control Item (∇) Requirements
- Engineering Specification (ES) Test Performance Requirements
- Prototype Part Quality Initiatives
- Advanced Product Quality Planning Status Reporting Guidelines, Ford Automotive Operations
- Run at Rate
- Supplier Laboratory Requirements and Calibration Services

- Table A—Qualification of All Product Characteristics
- Table B—Ongoing Process and Product Monitoring
- Ford Glossary

Tooling and equipment suppliers should reference the manual, *Ford's Failure Mode & Effects Analysis Handbook Supplement.*

General Motors-specific Requirements

The following sections of QS-9000 do not apply to tooling and equipment suppliers and are only for reference.

- Third Party Registration Requirements
- General Procedures and Other Requirements
- QS-9000 Applicability
- UPC Labeling for Commercial Service Applications
- Layout Inspection and Functional Test
- Customer Signatures on Control Plan
- GM Holdens-Specific Requirements
- PPAP
- Shipment Notification System—4.15.6.4

Tooling and equipment suppliers should reference the following manuals:

- *General Motors Corporation C4 Technology Program, GM—Supplier Information, (GM1825)*
  Assists suppliers in understanding and executing the GM C4 Strategy, and provides Year 2000 readiness information

- *All GM-specific requirements (GM 9000)*, referenced in the GM-specific section of the QS-9000

- *General Motors Corporation Vibration Standard for New and Rebuilt Machinery and Equipment*
  GM Specification No. V 1.0-1993 (or latest version) [GM-1761]

17

- *General Motors Corporation Laser Alignment Specification for New and Rebuilt Machinery and Equipment*
  GM Specification No. A 1.0-1993 (or latest version) [GM-1907]

Electrical	*SAE*	*HS-1738*
Electric Motor	*7EH*	*GM-1726*
Hydraulics	*HS1*	*GM-1744*
Lubrication	*LS1*	*GM-1720*
Pneumatic	*PS1*	*GM-1775*
Sound Level	*SL1*	*GM-1619*

Other OEM-specific Requirements

This section of QS-9000 does not apply to tooling and equipment suppliers and is only for reference.

A review of these requirements shows that the both the QS-9000 TE supplier and the customer have many responsibilities and fulfillments. The requirements depicted here have been abbreviated from the full requirements specified in the *TE Supplement*. However, what is presented here should give readers an adequate idea of what is needed. Keep in mind that the *TE Supplement* and the *R&M Guideline* are the documents to be followed and that QS-9000 must be implemented in its entirety except for the requirements defined by the *TE Supplement*. The *TE Supplement* specified requirements are in addition to QS-9000.

In March 1998, the third edition of the *Quality System Requirements QS-9000* was published. The second edition of *Quality System Requirements—Tooling and Equipment Supplement* was published in June 1998. The information shown here reflects both new editions. The Big Three revisions to the QS-9000 contain the following basic changes:

- Section II: Sector-specific Requirements has been included into the 20 ISO 9000-based elements (i.e., Section II in the third edition is now Customer-specific Requirements).

- Many of the International Automotive Section Group-sanctioned QS-9000 interpretations have been included in the 20 elements.

- The automotive additions are now numbered.

- Several European automotive requirements have been added.

- The glossary has been expanded.

The second edition of *Quality System Requirements QS-9000* became obsolete on January 1, 1999. Individual customers specify which revision is required for their suppliers.

References

Specific company documents that are specified in this chapter are available from the organizations that require them.

Quality System Requirements Tooling and Equipment Supplement, 2nd ed., Chrysler Corporation, Ford Motor Company, and General Motors Corporation, 1998.

Quality System Requirements QS-9000, 3rd ed., Chrysler Corporation, Ford Motor Company, and General Motors Corporation, 1998.

Reliability and Maintainability Guideline for Manufacturing Machinery and Equipment, Society of Automotive Engineers, Warrendale PA, and National Center for Manufacturing Sciences, Ann Arbor, MI, 1999.

Chapter 2

Basic Elements of Reliability and Maintainability

Reliability is defined as the probability that an item will perform its intended function for a specified interval of time (or some other measure) and in stated environmental conditions. In other words, the definition forces the designer to admit that the possibility of failure exists, that performance may deteriorate with time, and appropriate environmental and operational conditions exist under which a product will operate.

Maintainability is a characteristic of design, installation, and operation, usually expressed as the probability that a machine can be retained in, or restored to, specified operable condition within a specified interval of time when maintenance is performed in accordance with prescribed procedures. More simply stated, maintenance personnel must have the necessary skills, procedures, and resources to accomplish the task of restoration. The initial design can have a very high reliability when accompanied with specific maintenance procedures; however, if the procedures are not followed, there is no question that the reliability will degrade. The design engineer must assure that appropriate product maintenance requirements are specified. Maintainability must be a part and parcel of the design. However, as in manufacturing, if the parts are not built to specification or if they are not properly maintained as required, the reliability will suffer. Chapter 3 provides additional insight and details regarding maintainability.

With these definitions of reliability and maintainability (R&M), how does the designer design a product that meets the specific reliability requirements of a customer? Before the designer does this, several questions must be answered:

21

1. Is there a customer contractual or management reliability requirement that must be met?

2. Are there tradeoffs that must be considered with other design parameters?

3. Will the required reliability be cost effective? (Higher reliability does not always mean higher cost.)

4. Does the designer understand all of the conditions that the product will undergo and how it will be utilized?

5. Will the designer receive support and commitment from management for improving reliability if necessary?

6. Does the designer contemplate any manufacturing problems that may degrade reliability?

In 1996, Chrysler Corporation, Ford Motor Company, and General Motors Corporation published *Quality System Requirements—Tooling and Equipment Supplement*, hereafter referred at the *TE Supplement* or QS-9000TE. The *TE Supplement* was developed to specify the quality system requirements for suppliers desiring to become certified to QS-9000TE or to establish in-house operations so that the suppliers are in compliance. The intent of the *TE Supplement* is to:

- Provide an interpretation of the QS-9000TE requirements

- Make commonplace individual company system requirements with regard to R&M

- Provide focus in the application of process-driven quality systems

- Promote effective use of the principles of R&M

The *Quality System Requirements T&E Supplement* by the Big Three specifies the requirements that tooling and equipment manufacturers must meet. The concepts of Mean Time To Failure (MTTF), Mean Time Between Failures (MTBF), Mean Time To Repair (MTTR), and other R&M terminology must be understood and demonstrated as requested by the customer.

Many of us marvel at the wonders of science, but we are taken aback when we cannot take care of nuisance malfunctions such as a car failing to start and the breakdown of production machinery and appliances. For some products, competition forces a company to improve reliability; otherwise, that company will not be able to remain in business. This chapter is not a course in reliability, but it attempts to present an overview of reliability and maintainability and information about what they involve. Reliability thinking starts in the concept stage of a product. It should be a part of the design and should continue through development and production. On the other hand, maintainability provides a measure of repairability of a system where a certain amount of failures can be allowed. QS-9000TE suppliers are required to demonstrate specific Mean Time To Repair (MTTR) when assigned by the customer. As for reliability, the demonstration of these requirements must be met within stated conditions.

The Probability Theorems

The basic elements of reliability and maintainability include the use many probability distributions, theorems, predictions, tools, and techniques. These are utilized in various ways to determine the reliability and maintainability of a particular part, assembly, or system. Introductory descriptions are given in the following paragraphs.

To comprehend the meaning of reliability and maintainability, probability theory first must be understood. Fundamental theorems of probability usually are the foundation for calculating reliability and maintainability. These include:

Rule 1: The probability of an event lies between 0 and 1. Zero is the probability that an event will not occur, and one is the positive certainty that an event will occur.

Rule 2: The sum of the probabilities of a situation is equal to 1.00.

Rule 3: The Complementary Law states that if the probability of the event is P, the probability of the event not occurring is $(1 - P)$.

Rule 4: The Additive Law states that the probability of either A or B occurring is the addition of the two events. However, for this to happen, the two events must be mutually exclusive.

Rule 5: The Multiplication Law of probability states that the mutual probability of two independent events is equal to the product of the probabilities of each event.

Rule 6: The Combination Law is the combining of the additive and multiplication laws. It is used to find the probability of occurrence of either one or two events.

Rule 7: The Conditional Law applies to events that are dependent on each other, or

$$P(A \text{ and } B) = P(A) \times P\left(\frac{B}{A}\right)$$

Here, $\frac{B}{A}$ usually is referred to as "B given A."

These rules are explained in much more depth in textbooks on probability theory. They are presented here only to show the rules of probability. Examples of how these rules are used are demonstrated in other applicable sections of this text.

The Probability Distributions

The probability distributions also relate to reliability and maintainability. These are depicted and described in Fig. 2.1. The more familiar ones are shown here, with a brief explanation of how they are utilized.

The distribution curves in Fig. 2.1 show the different models to which R&M can relate. Other probability distributions are available, but these are the most basic ones. The use of these major distributions is utilized in R&M and is explained further in subsequent chapters of this book.

The probability distributions relate to the values that fall within the bounds of probability shown. If the values can take any value, the distribution is called a

DISTRIBUTION	FORM	PROBABILITY FUNCTION	COMMENTS ON APPLICATION
NORMAL		$y = \dfrac{1}{\sigma\sqrt{2\pi}} e^{-\frac{(X-\mu)^2}{2\sigma^2}}$ μ = Mean σ = Standard deviation	Applicable when there is a concentration of observations about the average and it is equally likely that observations will occur above and below the average. Variation in observations is usually the result of many small causes.
EXPONENTIAL		$y = \dfrac{1}{\mu} e^{-\frac{x}{\mu}}$	Applicable when it is likely that more observations will occur below the average than above.
WEIBULL		$y = \alpha\beta\,(X-\gamma)^{\beta-1} e^{-\alpha(X-\gamma)^\beta}$ α = Scale parameter β = Shape parameter γ = Location parameter	Applicable in describing a wide variety of patterns of variation, including departures from the normal and exponential.
POISSON*		$y = \dfrac{(np)^r e^{-np}}{r!}$ n = Number of trials r = Number of occurrences p = Probability of occurrence	Same as binomial but particularly applicable when there are many opportunities for occurrence of an event, but a low probability (less than 0.10) on each trial.
BINOMIAL*		$y = \dfrac{n!}{r!(n-r)!} p^r q^{n-r}$ n Number of trials r = Number of occurrences p = Probability of occurrence q = 1-p	Applicable in defining the probability of r occurrences in n trials of an event which has a probability of occurrence of p on each trial.
NEGATIVE BINOMIAL*		$y = \dfrac{(r+s-1)!}{(r-1)!(s!)} p^r q^s$ r = Number of occurrences s = Difference between number of trials and number of occurrences p = probability of occurrence q = 1-p	Applicable in defining the probability that r occurrences will require a total of r + s trials of an event which has a probability of occurrence of p on each trial. (Note that the total number of trials n is r + s.)
HYPERGEOMETRIC*		$y = \dfrac{\dbinom{d}{r}\dbinom{N-d}{n-r}}{\dbinom{N}{n}}$	Applicable in defining the probability of r occurrences in n trials of an event when there are a total of d occurrences in a population of N.

Fig. 2.1. Probability distribution curves. (From Juran's Quality Handbook, *5th ed., by Joseph M. Juran).*

"continuous" probability distribution. The normal, exponential, and Weibull distributions are good examples. When the characteristics can be only specific values, the distribution is called a "discrete" probability distribution. Examples are the Poisson, binomial, and hypergeometric distributions.

Fundamental Reliability Relationships

Component reliabilities as they relate to assemblies have the following basic relationships. For clarification, a legend follows the equations presented here.

- Components in Series (Product Rate)
  An extension of the Product Rule (equal component reliabilities)

$$R_s = \prod_{i=1}^{n} R_i$$

$$R_s = R_1 \times R_2 \times R_3 \times \ldots R_n$$

or diagrammatically as

$$-\boxed{R_1}-\boxed{R_2}-\boxed{R_3}- \cdots -\boxed{R_n}-$$

- Components in Parallel

$$R_s = 1 - (U_1 \times U_2 \times U_3 \times \ldots U_n)$$

$$R_s = 1 - \left[(1 - R_1)(1 - R_2)(1 - R_3)\ldots(1 - R_n) \right]$$

also

$$R_s = R_1 R_2 + R_1(1 - R_2) + R_2(1 - R_1) \quad \text{(for two components)}$$

or diagrammatically as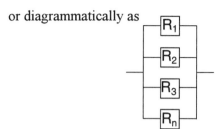

- Combined Series/Parallel Systems

$$R_s = R_1 \times R_2 \times \left[1 - (1 - R_3)(1 - R_4)...(1 - R_n)\right]$$

or diagrammatically as

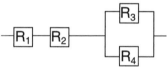

- Exponential Reliability

$$R = \exp\left(\frac{-t}{\theta}\right) \quad \text{or} \quad R = e^{-\left(\frac{t}{\theta}\right)}$$

- Reliability Using Weibull Distribution*

$$R = \exp(-\lambda t)\exp\beta \quad \text{or} \quad R = e^{-\left(\frac{t}{\theta}\right)^\beta}$$

---

* Chapter 7 provides instructions for calculation of the Weibull graph.

Legend for the equations:

$$R_s \quad = \quad \text{System reliability}$$

$$R_1, R_2, R_3 ... R_n \quad = \quad \text{Component reliabilities}$$

$$U_1, U_2, U_3 ... U_n \quad = \quad \text{Probabilities of failure for each component}$$

$$e \quad = \quad \text{Natural log} = 2.71828 ...$$

$t \quad = \quad$ Specification—time, miles, cycles, or usage—by which reliability is measured

$\theta \quad = \quad$ Mean life or average time of usage (can also be MTTF, MMTF, MTBF, etc.)

$$\lambda \quad = \quad \text{Failure rate} = \frac{1}{\theta}$$

$$\beta \quad = \quad \text{Weibull slope}$$

Subsequent sections of this chapter further explain these relationships.

Reliability Design Reviews

Reliability design reviews are formal reliability examinations of a product at key milestones in the development of that product. The reviews should start in the early design concept stage and continue through actual customer usage. The purposes of the reviews are to make disciplined searches for potential design problems, to identify possible alternatives, and to resolve potential problems that may arise. Members of the design review team should include the collective experience and judgment of those representatives who are involved in the design, manufacturing, and usage of the product. This usually includes personnel from engineering, marketing, quality, manufacturing, reliability, maintainability, testing, and any other discipline that can contribute to the improvement of the design. Chapter 4 provides more detail on design reviews.

Design Checklists

The use of design checklists is another method to assure and increase reliability in design. Several reliability textbooks have recommended checklists for the designer to use. A brief checklist is shown in Chapter 4. It is important for the designer is to ask himself or herself the types of questions that will ensure that all possibilities are covered. In some respects, the checklists are similar to a Failure Modes and Effects Analysis (FMEA) because the designer is forced to consider all the different possibilities associated with the product and how the product possibly could fail. The engineer can make his own checklist or use appropriate established ones.

Reliability Testing

Reliability testing is essential to the determination of product reliability. Without some form of testing, the engineer has no means of knowing whether the product meets all of the criteria of functional requirements. Testing can be done for prototypes, qualification, or acceptance. Testing is critical to any reliability program. Disclosure of deficiencies in the design and the verification of corrective actions to prevent any recurrence of failures are keys to a reliability program. Testing must relate to a planned test program, the Failure Reporting, Analysis, and Corrective Action System (FRACAS), and reliability growth and demonstration. The concept of Test-Analyze-And-Fix (TAAF) is utilized by most concerns; however, an even better concept is Analyze-Test-Analyze-And-Fix (ATAAF). This asserts that analysts on the design can be implemented early in the design stage to minimize test failures. Testing also helps in establishing reliability values that may be necessary for contractual purposes or for target goals established by the management of the concern. If the product will be used in extreme environments (e.g., those environments with high temperature and pressure, fungus, sand, dust, or explosive atmosphere), that product should be tested with those extremes. If vibration and accelerations are factors, these also should be considered.

Failure Modes and Effects Analysis/Fault Tree Analysis

Several techniques can be used to determine and measure reliability. One of the most rudimentary techniques is Failure Modes and Effects Analysis (FMEA), which is sometimes referred to simply as Failure Mode Effects Analysis (FMEA) or expanded to Failure Modes Effects and Criticality Analysis (FMECA). This technique provides a disciplined procedure whereby engineers can assess all possible failure modes that a product can possess. With this approach, the design or reliability engineer is forced to appraise all possible failure modes and then consider the probability of failure occurrence and the effect of each failure. All potential problems should be considered by identifying the cause and effect of each failure mode. These then are prioritized according to frequency of occurrence, severity, and detection. They are considered for corrective action if there is a need to improve the design. As more information about the design is learned, the FMEA is updated. Remember that the success of any product is being able to foresee the potential failures of the product and then being able to do something about them. Chapter 5 provides an FMEA format and discusses how it is developed.

Chapter 5 also discusses Fault Tree Analysis (FTA). The FTA can be thought of as a rearward or upside-down FMEA because the logic proceeds from the failure effects of interest to discover the causes, rather than from all the causes of possible failures to discover conceivable effects.

Reliability and Maintainability Predictions and Allocation

Reliability and maintainability (R&M) predictions and allocations are quantitative assessments of reliability and maintainability. They are key elements of formalized reliability maintainability planning and control. The techniques force the engineer to predict the reliability and maintenance parameters of the design. An extension of the reliability and maintainability status can be continued to the subsystems and component levels as well as the overall design. This information then is utilized to assess whether improvements are essential for specific subsystems or to determine if any tradeoffs are plausible. The allocation part of this technique forces the designer to apportion components and subsystems to some reliability and maintenance ratio. In turn, these are compared to the predictions and may necessitate some tradeoffs.

A U.S. astronaut once stated that the most nerve-wracking part of any space flight is the fact that there are thousands of critical parts in a space shuttle, each of which is produced by the lowest bidder. Most industrial products do not have this level of criticality, but the designer should understand the importance of probabilities that make up the assemblies. Overall, reliability is enhanced by minimizing the probabilities of malfunction of all components making up the system. Chapter 10 shows a simple schematic of a probability breakdown of a machine assembly and additional details on the prediction process. As a general rule, with everything else being equal, fewer parts in an assembly usually means a greater chance that the system has a higher reliability.

Data Collection

The analysis of data is a field in itself and relates closely to the determination of reliability, its analysis, and maintainability. The use of "go, no-go" or attribute data, tests to failure, variable data, different distribution patterns (including the normal, exponential, and Weibull, among many), design of experiments, correlation and regression analyses, and many other techniques all play an important role in reliability. The important concern is that information from test data must be valid and understood, and it must be recorded and accumulated in such a manner that it can be analyzed for reliability and maintainability purposes. Chapter 8 presents some of the methods of data collection and shows how data collection can be utilized in R&M.

Failure Analysis/Failure Reporting, Analysis, and Corrective Action System

A Failure Reporting, Analysis, and Corrective Action System (FRACAS) is one of the most important processes in an R&M program. Designs are drawn and analyzed, prototypes are fabricated and tested, and the test results are observed. Testing in one form or another is the only real way a product can be verified. To establish a FRACAS-type system, some formalized reporting of test results must exist. Test failures must be analyzed for root cause, and then corrective action must be implemented. Failure analysis of failed parts is critical to any R&M program and cannot be overemphasized. If the analysis of potential failure modes was accomplished thoroughly and early in the program, failure incidents should be at a minimum. However, in most cases, the design still must be tested to the conditions utilized by the customer. All corrective actions that are resolved into the design also must be verified through testing. The bottom line is that when a test malfunction occurs, the designer should know why and how the failure occurred and should be able to determine its root cause. Then the designer must implement corrective action to assure that the failure does not recur. Chapters 8 and 9 provide further details on FRACAS and failure analysis.

Field Reporting

Field reporting of failures is one method of data collection. It is the concluding indicator of how good a job the designer did. The final product is now in the customer's hands. Field reporting can indicate if manufacturing is producing the product according to specifications. Field reporting can exist in several forms, including:

1. Actual reported failures of individual customer usage
2. Warranty and extended warranty reports
3. Field service reports
4. Formalized reporting systems from the customer
5. Informal reporting such as complaint letters

Reliability testing theoretically is supposed to catch potential design problems. During the development stages of the product, reliability testing should be occurring (i.e., early prototype, qualification, acceptance, and then production configuration). However, the real "proof of the pudding" does not occur until the customer uses the product. The earlier that malfunctions can be found in the design phase, the greater the chance of success in the final product. Ultimately, the customer is the final user who utilizes the product as he or she thinks it should be used. This information can prove very valuable to the designer, especially if all of the potential user modes and environmental conditions were not considered during the design phase. Also, remember that costs can increase exponentially as the discovery of malfunctions occurs further into production and as the product gets into the customer's hands. Chapters 7 and 8 provides additional details on field reporting.

Reliability Growth, Measurement, and Demonstration

Reliability growth, measurement, and demonstration are essential for the reliability determination of a product. The techniques relate closely to reliability testing. Test planning and objectives are important considerations as a part of a development program. The idea is to take a concept and transform it to a reliable product. Techniques for testing vary, from bench tests to testing a one-shot, full-scale product such as a missile. Between those extremes, numerous arrays of test methods exist. These include testing to failure to determine lives of individual components in a product; testing to specific miles, hours, or cycles to determine if a standard can be met; accelerated testing; testing for infant mortality; testing for wear-out; and testing to determine random failures. Chapter 11 presents ways in which reliability growth can be measured and the manner in which reliability demonstrations can be depicted.

Improving Reliability

Some of the methods the designer uses to improve reliability are basic; some methods are complex. The primary thought is to keep the design as simple as possible, uncluttered and uncomplicated. Of course, this cannot always be accomplished because of the nature of some products. However, it is something for which the designer should strive.

The designer should utilize proven margins and confirmed components whenever possible. It is always convenient to use component designs that have had a good history of performance and to use specifications that correlated well with function. Although one of the concerns to which designers should be alert is the use of proven factors in the new product, keep in mind that the components may undergo different environments or customer usage. "Keep It Simple, Stupid" (KISS) is a good adage to remember when designing a product. Another similar term is "idiot proofing," which means to make the product so that a person does not make mistakes in using or producing that product. Engineers should keep these principles in mind whenever possible. Chapter 14 presents other details and methods on the way reliability can be improved.

Redundancy

Redundancy is the existence of providing alternate means of accomplishing a given function in the event that one of the components within the system fails. Redundancy can be extremely expensive and can add extra weight if it is used on many assemblies in a system. Redundancy is especially useful in electronic systems where relatively low-cost, lightweight components can be connected together in parallel. If the reliability requirement of the subsystem is high, the designer may have no choice but to use redundancy to attain the high value.

Human Factors

Human factors are important considerations in the design of the product. Questions such as the following should be asked:

1. What are the possible ways in which human error can disrupt the system or cause it to fail?

2. Are parts and operations "idiot proof"?

3. Are operating controls and levers within the reach and control of specified personnel percentiles?

4. Are operators and maintenance personnel properly and adequately trained to perform their responsible functions?

5. Are instructions adequate and clear?

6. Is adequate space allowed for the performance of tasks?

7. Is adequate time allocated for the completion of tasks?

8. Have temperature, humidity, and other environmental conditions as related to humans been considered?

9. Have man-to-man and man-to-machine interfaces been considered thoroughly?

10. Have all safety hazards been considered? (This may be the last item listed here, but it is extremely important.)

If humans will be interfacing with a product, all of these factors should be considered. Many of the questions can be answered quickly, but others may require a great deal of thought.

Benchmarking

Another tool that can be utilized when considering reliability is benchmarking, which is the searching for best practices that can lead to a superior product or performance. This concept is not new. In the past, it may have been viewed as looking at the competition and then trying to do something better than what is being done by the competition. Benchmarking goes beyond merely looking at and amassing information about a competitor. It looks continually at the ways "best-in-class" organizations operate, and then it attempts to further improve what is observed.

As with other quality tools, benchmarking requires serious commitment at all levels of an organization to yield continuous improvement performance. Particulars such as product and service enhancements, achievement indicators, work processes, and organizational strategies are a few areas to which benchmarking can apply. The intent is to uncover the best practices wherever they exist, and then to enhance them further. Benchmarking should be a proactive activity rather than a reactive one. Organizations should invariably search for positive changes and strive for superior performance.

From a reliability point of view, engineers should focus on competitive and best-in-class designs and processes. They should seek out functional experts and networking professional associates, review trade publications, and access service agencies. Internal engineering operations usually are accessible within an organization where engineers can work in unison to improve on their designs. This approach may be a little difficult with competitors, but site visits to facilities other than those in an engineer's own industry often can provide advantageous comparisons and useful information. The idea of benchmarking is to be innovative in terms of implementation and application, rather than be an imitator or a part of a catch-up exercise. It can be a great tool in the leading edge of setting reliability goals that often relate with market-driven necessities.

Safety

Safety is one of the elements under human factors. The subject is important enough to stand alone. In other words, there should be at minimum an assessment of the design to assure compliance with all safety codes and that no safety concerns exist. Then, as in reliability testing, the design must be verified to assure that all safety aspects of the product are identified and that the design meets all safety and hazard criteria. Maintainability procedures also should take into account the safety and hazard of a product. Maximum attainable safety can be realized only by knowing the reliability of the equipment components. Chapter 16 provides additional information on safety and how it relates to reliability.

Bathtub Curve

One of the most classic features about reliability is the bathtub curve. The curve, shown and described in Fig. 2.2, characteristically describes the life of many products. The curve originated from insurance companies that required some measure of actuarial rates. In the 1960s, reliability engineers also saw the characteristics of the curve as being applicable to product life. In effect, the curve outlines the early infant mortality-type failures in a product, then an array of random failures, and then wear-out. The bathtub curve tells much about a product. Infant mortality failures can be reduced by incorporating good quality control practices; by having preliminary burn-in, debugging, environment stress screening, or pre-delivery tests; or by using highly reliable components that do not display infant mortality-type failures.

The random incidents usually are failures for which the root causes are difficult to determine. A reliability engineer may claim that there is no such thing as a random failure, but root cause at times can be difficult, if not impossible, to determine. Of course, this depends on the circumstances. Random failures can be the isolated case, the one-in-a-million scenario (but on different components within the product), such as the one bolt that was not Loctited or a solder joint that was not manufactured to specification.

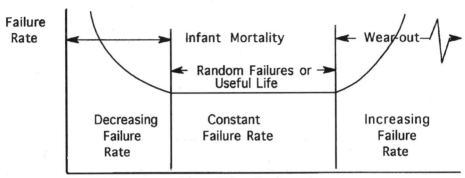

Fig. 2.2. Reliability bathtub curve.

1. *Infant Mortality: Period characterized by high failure rates that appear in early usage. These failures usually stem from quality or manufacturing problems, immature design, misapplications, or other types of causes.*
2. *Constant Failure Rate: Period of random failures, marginal design, poor maintenance, or misuse.*
3. *Wear-Out: Period in which components start wearing out, usually as a result of "old age."*

The wear-out portion of the curve represents those parts that begin to show the limits of their endurance. Most parts have a finite life and at some point will deteriorate slowly from prolonged use. From a design perspective, this wear-out feature must be out far enough in the life cycle of the product so that customers at least feel they bought a reasonable bargain. This is especially true if the product was maintained properly and not abused. Competition, reputation, cost of the product, and other factors determine what kind of bathtub curve a company can tolerate. A product that has little or no infant mortality, a very low constant failure rate, and a wear-out portion that is far out in its life cycle will certainly sell much better than a product lacking these features, sometimes even if the price is higher. Reliability then becomes a tradeoff between cost, competition, and the desire to satisfy the customer.

Other techniques can be useful in analyzing data and in determining and improving reliability. These include the use of Pareto charts, histograms, probability theory, confidence intervals, supplier control and monitoring, different distribution patterns (especially Weibull graphic analysis), cause-and-effect diagrams, scatter diagrams, and design of experiment. Chapter 15 provides additional details of the basic "problem solving tool." Other chapters discuss some of the alternative methods. Chapter 18 describes methods for administering and managing R&M, as well as the benefits and advantages of having an R&M program. All of these are important elements of R&M and must be understood to a meaningful degree in order to establish and maintain a valid R&M program. The following chapters present more details on these techniques. The chapters also introduce other topics essential for meeting the QS-9000TE requirements.

References

Dhillon, Balvir S., and Reiche, Hans, *Reliability and Maintainability Management*, Van Nostrand Reinhold Co., New York, 1985.

Doty, Leonard A., *Reliability for the Technologies*, 2nd ed., ASQC Quality Press Book by Industrial Press, New York, 1989.

Juran, Joseph M., *Juran's Quality Handbook*, 5th ed., McGraw-Hill, New York, 1999.

Juran, J.M., and Gryna, Frank M., *Quality Planning and Analysis*, 3rd ed., McGraw-Hill, New York, 1993.

Lewis, E.E., *Introduction to Reliability Engineering*, 2nd ed., John Wiley and Sons, New York, 1994.

O'Connor, Patrick D.T., *Practical Reliability Engineering*, 2nd ed., John Wiley and Sons, New York, 1985.

Chapter 3

Maintainability

In Chapter 2, we defined maintainability as a characteristic of design, installation, and operation, usually expressed as the probability that a machine can be retained in, or restored to, a specified operable condition within a specified interval of time when maintenance is performed in accordance with prescribed procedures. Another way of defining maintainability is the probability that a failed system can be restored to operating conditions within a specific interval of downtime. Maintainability parameters are reported in formats such as Mean Time To Repair (MTTR), Maintenance Ratio (MR), Mean Time Between Unscheduled Maintenance Actions (MTBUM), and a multitude of other attributes and descriptions.

Factors to keep in mind when formulating maintainability are as follows:

- The adequacy of the maintenance procedures with regard to the product

- The environment under which the maintenance is performed

- Adequacy of personnel with required skill levels

- The equipment design and installation with regard to maintainability

- Required maintenance time to keep a product operational

- Unscheduled maintenance time to restore a service in the event of a failure

- Providing input to the design process to facilitate fault identification

- Reducing overall Life Cycle Cost (LCC)

One of the main concerns of a reliability engineer (or a maintenance engineer, depending on the size and structure of the organization) is to consider maintainability as an integral part of the design. The idea is to reduce the number of failures and, where possible, apply the tools of preventive maintenance while at the same time restoring a product to its operative condition as soon as required (i.e., corrective maintenance).

One of the measures of maintainability is the MTTR, which indicates the mean time required to repair a component, assuming the spare parts and technicians are available to perform the work. Similar to reliability, there is no one maintainability index that applies to most products. If the customer stipulates a specific requirement, the QS-9000TE supplier is obligated to assure that the specified maintainability index for the product is as designated.

Other maintainability requirements that may be specified are the Maintenance Manual, Maintenance Ratio (MR), percent of downtime due to hardware failures or software errors, Mean Time Between Unscheduled Maintenance (MTBUM), and Mean Time to Perform Preventive Maintenance (MTPPM). The customer also may specify other indexes. The following equations can be used for calculating these characteristics.

Mean Time To Repair (MTTR) is calculated by developing the sum of the repair times by the number of repairs

$$MTTR = \frac{(\Sigma t)}{n}$$

where

$$MTTR = \text{Average time to repair an item}$$
$$\Sigma t = \text{Sum of individual repair times}$$
$$n = \text{Number of repairs}$$

The Maintenance Ratio (MR) is calculated by dividing the sum of maintenance man-hours divided by the total operating hours or some other cumulative index specified by the customer.

$$MR = \frac{MMH}{OH}$$

where

$$MR \ = \ \text{Maintenance Ratio}$$
$$MMH \ = \ \text{Cumulative maintenance man-hours}$$
$$OH \ = \ \text{Cumulative operating time, miles, cycles*}$$

Mean Time to Perform Preventive Maintenance (MTPPM) is calculated by dividing the sum of preventive maintenance hours by the number of preventive maintenance actions. This index sometimes is used as a measure of system effectiveness.

$$MTPPM = \frac{PMMH}{PMH}$$

where

$$PMMH \ = \ \text{Sum of the preventive maintence man-hours}$$
$$PMH \ = \ \text{Number of preventive maintenance actions}$$

Mean Time Between Unscheduled Maintenance (MTBUM) is another index used to measure system effectiveness and is calculated by dividing the total time of unscheduled maintenance actions by the number of unscheduled actions.

$$MTBUM = \frac{UMMH}{NUA}$$

---

* The customer defines the parameters.

where

$$MTBUM = \text{Mean Time Between Unscheduled}$$
$$\text{Maintenance activity}$$
$$UMMH = \text{Total of unscheduled maintenance man-hours}$$
$$NUA = \text{Number of unscheduled maintenance activity}$$

Another maintainability characteristic is Availability (A), which is defined as a measure of the degree to which machinery/equipment is in an operable and committable state at any point in time. Specifically, it is the percent of time that machinery/equipment will be operable when needed. Equations that relate to availability include

$$A_o = \frac{MTBF}{MTBF + MDT} \quad \text{and} \quad A_i = \frac{MTBF}{MTBF + MTTR}$$

where

A_o = Operational Availability when total downtime is used

A_i = Intrinsic Availability when active repair time is used

MTBF = Mean Time Between Failures

MTTR = Average time to repair an item

MDT = Mean downtime

The Maintenance Manual is a document that probably will be required by the customer. If the machinery is a simple design and does not require maintenance, a maintenance manual is not required. However, some kind of documentation should exist to specify to the user how the equipment should be maintained. When required, the maintenance manual usually includes comprehensive and detailed instructions, illustrations, periodic maintenance actions, and information that specifies what should be done to a product to keep it in an operating mode. In addition to these requirements, the manual also may specify the levels of maintenance activity. These include areas (from the most basic to the return to the factory) during operations, on operable equipment, on site, bench work, to a branch facility, depot repair, and factory repair. The maintenance procedures

42

usually are outlined in a step-by-step approach. Description of the required skill levels and necessary tools also could be included in the manual. In effect, the Maintenance Manual is the document that provides the directions for doing the maintenance on the product.

Maintenance activity includes scheduled inspections, tests, and overhauls; scheduled servicing; and unscheduled servicing, such as repairs and replacements. In addition to periodic maintenance actions and repair instructions, other directions may be included in the Maintenance Manual. Sometimes the customer may specify a separate repair manual. Customers should utilize and follow the directions so that their products will operate as specified.

Factors that reliability engineers should consider when developing a design include some of the following approaches:

- Depending on the availability requirement of a product, should the design show more improvement in reliability or maintenance?

- Modular or non-modular construction can have a large impact on maintainability.

- Built-in or external test equipment can influence cost, where built-in usually is more expensive.

- Repair versus throwaway means that when a repair becomes so expensive in the field, a throwaway option (or complete replacement) may be the most economical choice.

- Consider person or machine dictates about whether the operation/ maintenance is part of a highly engineered function with special instruments, to be left to a skilled technician, or to be engineered so simply that almost anyone could operate and maintain the product.

These approaches all must be weighed and evaluated, depending on how the customer wants the product. Tradeoffs are a prevalent part of R&M, and design engineering usually must determine the best way of designing the customer's product.

For QS-9000TE suppliers, the maintainability requirement usually will be specified as MTTR; however, other indexes also can be stipulated. Suppliers and customers both should be aware of the environmental conditions and how the machinery will be used. Any usage outside specified limits can cause deterioration of reliability. It is also expected that any other maintenance procedures that are required should be performed as specified. Neglecting to follow specified instructions can negatively impact reliability.

After the initial design and in early testing, maintainability requirements can be demonstrated. In much the same manner as reliability demonstration, maintainability can be explained by measuring the time to repair nonconformances or to perform other maintenance tasks.

The *R&M Guideline* specifies the maintainability requirements by the life cycle phase in which they fall. In the concept stage, machinery users are responsible for specifying the MTTR or Mean Time To Replace (also MTTR) to the supplier. Both the customer and the manufacturer must recognize that the required MTTR may not be met if operations are conducted outside the stated conditions.

In the design and development phase, the supplier should assure that the required maintainability features are part of the design. Downtimes due to corrective actions should be kept to a minimum. During Phase 2, suppliers should consider the following characteristics that will influence maintainability:

- Standard tool and test equipment should be used, unless specified other wise by the customer.

- Manuals should be consistent with skill levels of maintenance personnel.

- Schedules and inventories should be compatible with preventive maintenance (PM) requirements.

- Time to complete PM procedures should be defined.

- Locations for PM analysis should be identified.

- Spare parts lists should be based on equipment reliability characteristics.

- Spare parts inventory should be kept to a minimum.

In the operation and support phase, users are expected to supply the TE manufacturers with all maintenance information so that any oversights or misjudgments can be corrected by the manufacturer. The importance of data collection in this stage cannot be overemphasized. The key to any successful R&M program is well-documented problems and concerns, with subsequent follow-up for corrective and preventive actions and implementation. The core element during evaluation and testing is to demonstrate that maintainability goals have been achieved. Users must understand the importance of failure data and maintenance information.

Many actions take place during maintenance activities for machinery. These include both corrective and preventive actions, sometimes in combination. The following list identifies some of the actions that prevail to keep machinery operating effectively, or restoring machinery to serviceable condition.

- Accommodating changing organizational needs
- Adjustments that must be made
- Administrative directives
- Alignments
- Assembly
- Calibration
- Checkouts
- Cleaning
- Conditional determinations
- Correcting
- Diagnosis
- Disassembling
- Engineering support
- Inspections
- Localization
- Lubrication
- Measurements
- Modifications
- Overhauling
- Preparing reports
- Rebuilding
- Reclamation
- Reinstallation
- Removal

- Repairing
- Replacement
- Restoration
- Securing equipment, supplies, and materials
- Testing
- Troubleshooting
- Trying out new and trial components

Chapter 10 discusses the R&M predictions that are required by QS-9000TE suppliers. The user defines the requirement parameters. The manufacturer then determines the maintainability predictions. This is the quantitative tool that the manufacturer believes it can attain and that the user deems acceptable.

In summary, we can see several characteristics that relate to the important measure of system effectiveness in maintainability. The adequacy of maintenance procedures and test equipment are important factors. The environment under which maintenance is performed should be as close as possible to that in which the user would be practicing. Machinery design and installation should meet the requirements of the customer, and consideration should be given to the skill level of the user's maintenance personnel.

References

Doty, Leonard A., *Reliability for the Technologies*, 2nd ed., ASQC Quality Press Book by Industrial Press, New York, 1989.

O'Connor, Patrick D.T., *Practical Reliability Engineering*, 2nd ed., John Wiley and Sons, New York, 1985.

Chapter 4

Design Review

Design review is the process in which involved individuals systematically study a design, or parts of a design, to optimize the R&M characteristics of a product. The reviews usually are formal, documented, methodical, comprehensive, and intended to satisfy the design requirements of the customer. The reviews can be considered an integral part of the design and development process, and they can extend well into the operation and support phases of a product.

Design reviews should start in the concept stage of a product and continue to production and usage by the customer. Knowledgeable individuals with the required skills and cognizance of the customer meet periodically to consider the fabric of the design and also to improve on what has been developed and manufactured in the past. These skill groups include marketing, safety, design and test engineering, and quality and manufacturing engineering. The intent is to at least meet the R&M requirements of the customer.

Design reviews are not new. They were done in the past somewhat informally and probably without much documentation, except for the designer taking a few notes about agreements and conclusions from the individual with whom he spoke. Because of the complexity of modern designs and the importance of involving more skills and disciplines, design reviews today are more formal and methodical. They also should document actions that have been taken or scheduled, and who is accountable for specific responsibilities. The formality of the design review recognizes that specialized knowledge is required in several disciplines and that the skills must work together to achieve an optimum design.

The following points should be considered with regard to design reviews. These points also can be used as a checklist.

- Design reviews must be conducted because of customer or certification requirements.

- Design reviews should be supported fully by management.

- Outside specialists often can be a part of the team (either temporarily or permanently) to utilize their experiences and knowledge.

- Design reviews should cover all aspects of the design, such as R&M, producibility, safety, LCC, weight, customer packaging and labeling requirements, appearance, and any other characteristics required by the customer.

- Design reviews are planned and scheduled during key points in the development process.

- Design reviews are documented with dates, attendees, action items, agreements, etc.

- The final decision on the design is the responsibility of the designer (with some official or unofficial input from management), who takes into account the knowledge and concerns from all other participants.

Designers classically believe that completion of the design is the end of the designer's action. This may be true to some extent, but the designer must realize that many factors are part of the design. The designer must think about the manufacturing, testing, and utilization of the product. He or she also must understand and appreciate that input is required from many individuals.

The design review team considers factors such as cost, reliability, maintainability, performance, producibility, size, weight, environment, and schedules. Both the machine supplier and the user review the technical aspects of the evolving design. The process generally includes the review of concept ideas, early sketches, drawings, standards, engineering notes, analysis determinations, test documentation and results, mockups, assemblies, hardware and software concerns, and the customer requirements.

The *R&M Guideline* presents a Design Review Objectives matrix that cross-links the design phase with the design review objectives. Table 4.1 shows the review aspirations and purposes at the various program phases.

TABLE 4.1
DESIGN REVIEW OBJECTIVES

Design Phase	Review Objectives
1. Concept	• Concept Review: Focus on the feasibility of the proposed design approach.
2. Development and Design	• Preliminary Design Review: Validate the capability of the evolving design to meet all technical requirements.
3. Build and Install	• Build: Address issues resulting from machine build and runoff testing.
	• In-Plant Installation: Conduct failure investigation of problem areas for continuous improvement.

Table 4.1 shows that the design review process must have many phases and be involved in all of the different design phases. In fact, the reviews also should extend to the operations and support phases. Feedback of customer information often reveals factors that may have been forgotten or considered too lightly to be of importance. The "proof of the pudding" is when the product is in the customer's hands. If design reviews are not taken seriously in the sequential design phases, warranty costs can exceed early budgetary considerations.

The *R&M Guideline* recommends three main areas when design reviews should be conducted (see Table 4.1 in the *R&M Guideline*). Some "reliability gurus" further extend the concept of design reviews by providing more details about additional phases that should be included. These are summarized here.

- **System Requirements Review:** This is the first review with the customer where the customer specifies the level of cost-effectiveness that he or she expects the manufacturer to meet. At this meeting(s), the customer and manufacturer reach agreement on the R&M characteristics and also the adjunct attributes of availability, dependability, and capability.

- **System Initial Design Review:** This meeting(s) provides the final definition of the system functional requirements, solidifying what was previously discussed. The allocation of R&M values accompanied by the attributes that support them usually is accomplished at this review.

- **Preliminary Design Review:** This is the point at which the configuration items are reviewed, and the complexities and technologies are discussed. The objectives here are: 1) To establish design adequacy and determine risks involved with the proposed design methods and techniques, 2) To harmonize the proposed design with the specifications, and 3) To ensure compatibility of the physical and functional characteristics with each other and with the operating and maintenance environments. Resolution of the entire system could be accomplished at this review, where there is some finalization on at least "first intentions." This would include initial sketches and drawings, mockups, simulation models, and prototypes.

- **Interim Design Review:** Formality of meetings continues by the manufacturer reviewing progress with the customer. Data feedback on all testing and analysis should continue as in previous reviews. The design starts to become more intense at this stage. The customer and supplier review the status of the milestones to ensure that the project is on target.

- **Critical Design Review:** By this stage, the design should be firm enough to "lock in" the design parameters and allow preparations to be made for qualification testing. The customer definitely should be present at this design review to ensure agreement on all particulars of the contract.

- **Formal Qualification Review:** This is the final review to precede full-scale production. The qualification models were fabricated and assembled with production tooling; however, at this stage, the customer and the manufacturer both should be ready for the production line. At this meeting, the review team confirms that the total package is as agreed and that it meets all terms of the contract.

- **In-house Reviews:** If problems occur after production begins, additional reviews may be necessary. Most R&M texts do not discuss this type of review because the design theoretically is "frozen" and the only problems, if any exist, are production problems. However, in the real world, anomalies occur and must be addressed. The design review concept should continue until the customer is totally convinced that he or she has a quality product.

Many questions arise during the course of design review meetings. Some organizations use a checklist to ensure that "nothing falls through the cracks" or that all potential problems are covered. Table 4.2 shows a generic checklist, which can be used at design review meetings or by the designer to ensure the integrity of the design. Tooling and equipment manufacturers and suppliers should review the checklist, eliminate any unnecessary items from the checklist, add to the checklist, or develop their own checklist. In any event, the checklist will provide something from which manufacturers can work.

TABLE 4.2
DESIGN REVIEW CHECKLIST

1. Do specified components meet their reliability requirements?
2. Can off-the-shelf items be used for particular functions?
3. Does the design meet functional requirements?
4. Were standard components and assemblies considered and used where possible?
5. Were all environmental impacts considered?
6. Did all components pass the environmental testing? Were corrections made where necessary?
7. Have critical characteristics been considered?
8. Have failure histories been investigated?
9. Did each component and material meet its requirements under environmental extremes of the specification?
10. Was there enough data for reliability calculations?
11. Was the complete unit tested?
12. Were the weak links in the design corrected?
13. Does demonstrated reliability meet the required specification? Or is redesign indicated?
14. Are predicted and allocated reliabilities compatible? Are tradeoffs necessary?
15. Were Manufacturing and Quality Assurance (QA) considered in the design?
16. Is redundancy necessary to meet reliability requirements?

TABLE 4.2 *(continued)*

17. Can the environment be changed or protected? Is heating, cooling, shock mounting, shielding, or better insulation required?

18. Were all failure modes corrected to prevent recurrence? Has storage capability been studied?

19. Should specifications be written to assure 100% test and inspection?

20. Are suitable manufacturing and QA procedures in place to assure good quality?

21. Is the item designed as simply as possible?

22. Have all human factors been considered? (See Chapter 2 for a list of human factors questions)

23. Have sharp corners been considered? Have all potential stress risers been eliminated?

24. Can other cognizant disciplines assist in writing specifications? Can they offer improvements in the design?

25. Can suppliers provide reliability values for their components? If so, are the values compatible with the overall system?

26. Has enough testing been performed to validate the required reliability for designated components?

27. Will early testing or screening help eliminate infant mortality-type failures?

28. Have maintainability requirements been considered?

Concurrent or simultaneous engineering is not a design review function, but it is closely related to it. Concurrent engineering is the process by which all disciplines work together to develop and produce a product. These include design, manufacturing, inspection, marketing, and maintainability. Milestones are established where the various disciplines accomplish specific tasks simultaneously before proceeding to the next task. In former times, each step in the design process occurred one step at a time and extended over a relatively long duration. In concurrent engineering, several tasks occur simultaneously, such as manufacturing engineering working with the quality engineer and design to set up their responsibilities while the designer is still working with the design. Table 4.3 shows the relationship between past conventional and concurrent methods.

TABLE 4.3
COMPARISON OF CONVENTIONAL AND
CONCURRENT ENGINEERING METHODS

Function	Conventional Engineering	Concurrent Engineering
Organization	Engineering is unique and separate from other departments	Engineering is part of a multifunctional team with team objectives
Timing of Outputs	Each department waits for output from previous departments	Tasks progress simultaneously working with adjunct functions
Design Change Frequency and Timing	Changes occur after testing or during production	Changes usually occur early, before the design is finalized
Information Systems	Delays of paperwork awaiting transfer of knowledge among functioning departments	Information is released as it is generated; functional departments have immediate access to information
Physical Location of Functions	Departments usually work in separate locations	Teams usually work together in one area

The advantage of concurrent engineering is that development time is considerably compressed and that the involved disciplines work together simultaneously. An enormous amount of rapport and understanding develops among the disciplines. In many cases, teams must be a part of the process to ensure that all disciplines and functions can fulfill their respective responsibilities. If project management wants the project to proceed on a timely basis, working together is essential. The end results usually are fewer engineering changes and a much faster time to market.

When examining the design review function, we can see that it is an essential part of the development and manufacturing process. Personnel in both the manufacturer's and customer's organizations must work in unison to ensure a reliable product. The design review process is one of the critical functions in which this can be accomplished.

References

Doty, Leonard A., *Reliability for the Technologies*, 2nd ed., ASQC Quality Press Book by Industrial Press, New York, 1989.

Feigenbaum, Armand V., *Total Quality Control*, 4th ed., McGraw-Hill, New York, 1991.

Jones, James V., *Engineering Design Reliability, Maintainability, and Testability*, Tab Professional and Reference Books, Blue Ridge Summit, PA, 1988.

Lloyd, David K., and Lipow, Myron, *Reliability: Management, Methods, and Mathematics*, Prentice-Hall, Englewood Cliffs, NJ, 1962.

O'Connor, Patrick D.T., *Practical Reliability Engineering*, 2nd ed., John Wiley and Sons, New York, 1985.

Chapter 5

Failure Modes and Effects Analysis and Fault Tree Analysis

Failure Modes and Effects Analysis (FMEA) is a technique originally developed in the automotive industry in the 1970s to identify potential failure modes and their effects on product performance. The technique frequently is used in reliability engineering to catch potential failures before they occur in the hands of the end users. In the QS-9000TE requirement, a complete FMEA starts in the concept stage of a program and spans the entire life cycle of the manufacturing equipment. Feedback from tester and user data collection points during the development of the product is essential to a good FMEA. The *R&M Guideline* suggests the following steps when utilizing the FMEA in a design of process evaluation:

1. Identify potential failure modes in the system and the expected seriousness of failure.

2. For each failure mode, study the effects on the total system.

3. Determine the potential cause and probability of occurrence of the anticipated failure.

4. Identify existing controls that are designed to prevent the failure from occurring without detection or to detect the failure and initiate corrective action.

5. Assign risk priority numbers to each failure mode, taking into account frequency of occurrence, severity, and likelihood of detecting the failure before it occurs.

6. Identify actions required to prevent, mitigate, or control failures; improve the likelihood of detecting the failure if it occurs; or eliminate the failure.

7. Assign an individual or group the responsibility of performing the recommended action. This includes documenting the action taken to implement the recommended action.

A block diagram can be constructed to systematically note the potential modes of failure and to quantify the likely parameters. Block diagrams may vary among users; an analyst could use any one of them appropriate for his or her needs. Figure 5.1 shows a typical layout for an FMEA, but it does not have to be exactly this way. The manufacturer can design its own layout or may have to use what the customer dictates. The *R&M Guideline* discusses the FMEA and provides instructions for its development; however, it does not recommend a specific format. QS-9000TE suppliers can fabricate their own formats, as long as they meet the requirements of the *TE Supplement*. For development of an FMEA, instructions are shown below and after Fig. 5.1.

These instructions and the FMEA format are applicable for product engineering, process manufacturing, or quality control analyses. The number in each step corresponds to the areas shown on the Fig. 5.1 FMEA worksheet.

1. Enter the subsystem or process name.

2. Enter the model number, part number, or the process identification (or whatever is applicable).

3. Enter the department responsible for the design or process.

4. Indicate if an outside supplier is involved.

5. Indicate the name of the subsystem engineer, design engineer, or process engineer.

6. Indicate the name of the system engineer.

7. Indicate the date on which the subsystem is scheduled for release.

Subsystem/Process ______(1)________ Process or Product Engineer____(5)___________

Model or Part No. ________(2)__________ System Engineer ____________(6)___________

Department Responsible _____(3)________ Sched. Product Release ________(7)____________

Outside Supplier __________(4)_________ FMEA Date Orig. ________ Rev. ______(8)__

Part/Process	Part Process Function	Failure Mode	Effects of Failure	Cause(s) of Failure				Risk Prior. No.	Recommended Corrective Action Action & Status
(9)	(10)	(11)	(12)	(13)	(14)	(15)	(16)	(17)	(18)

Fig. 5.1. Worksheet for Failure Modes and Effects Analysis (FMEA).

8. Show the date of the first FMEA completed on the product and then any subsequent dates of revision.

9. Specify the assembly, component, or process being analyzed.

10. Indicate the brief function of the part.

11. Describe each possible failure mode.

12. List all possible effects of the failure.

13. List all causes assignable to each failure mode.

14. On a scale of 1 to 10, evaluate the probability of occurrence. For example, a "1" indicates a very improbable occurrence, but a "10" indicates a very probable occurrence.

15. On a scale of 1 to 10, evaluate the severity or estimated consequence of the failure. For example, a "1" indicates a minor nuisance; a "10" indicates a several total failure.

16. On a scale of 1 to 10, estimate the probability that a potential problem will be detected before it reaches the customer. For example, a "1" indicates high probability that the failure would be detected before reaching the customer. A "10" indicates low probability that the failure would be detected before reaching the customer.

17. Multiply columns 14, 15, and 16. This provides an indicator of the relative priority of the failure mode (i.e., the higher the value, the more the concern). Some customers may require a reassessment of a particular failure mode if the value of this indicator is above a specific number. If that is the case, the manufacturer will have to review the failure mode and bring the value of concern to the level that the user requires.

18. Enter a brief description of the recommended corrective action.

Remember that the FMEA does not have to have the exact format presented here. Other arrangements are available, but each focuses on all possible failure modes and then arrives at a "probability-like" conclusion. The ultimate goal is to reduce as many failure modes as possible or to reduce the probability of the failure modes as much as possible.

Fault Tree Analysis (FTA) is an analytical technique developed by the aerospace industry in the 1950s and currently is used by other organizations as well. The technique utilized a "top-down" approach to failure analysis, starting with an undesirable "top event" and then determining all the ways in which the failure can happen. The analysis starts early in the concept phase and is monitored continually throughout the subsequent phases.

During the early 1950s when America was intent on developing a reliable intercontinental ballistic missile system, the U.S. Air Force realized the possibility existed that an inadvertent launch could create a serious safety problem. This realization prompted the Air Force to contract Bell Telephone Laboratories to develop an exact analytical technique that could be used to estimate the probabilities associated with an unacceptable event. Bell Telephone accepted the challenge and developed the FTA by the mid-1950s.

The FTA technique consists of the following and is still in general use today:

1. The qualitative and quantitative development of a block diagram that portrays a sequential array of factors that contribute to an inadvertent hazardous or unacceptable event.

2. The qualitative development is the logical analysis that identifies the group events, starting with the top event and then progressing downward to subevents of failures.

3. The quantitative approach utilized probabilistic mathematics to provide estimates that the undesirable event will occur. This included calculating failure probabilities for the various events.

4. Boolean algebra (sometimes called symbolic logic) is applied to formulate the deductive logic to the contributing fault-producing events that are related to each other.

The *R&M Guideline* defines the Fault Tree Analysis as a top-down approach to failure analysis, starting with an undesirable event and determining all the ways in which the event can happen. The first step in the analysis usually is to suppose that a failure has taken place or can take place. This is called the "top event." Then, the method considers all possible causes that could lead to the undesirable event. The analysis next looks for the potential origins of the failure causes and then considers ways to avoid these origins and causes. Figure 5.2 shows a simple FTA and light circuit, taken from the *R&M Guideline*. The figure depicts a light failure as the top event, followed by the possibilities of how the failure can occur with the subevents. (Incidentally, one other subevent can be added to the diagram, and that would be an inadvertent break in the circuit wiring.)

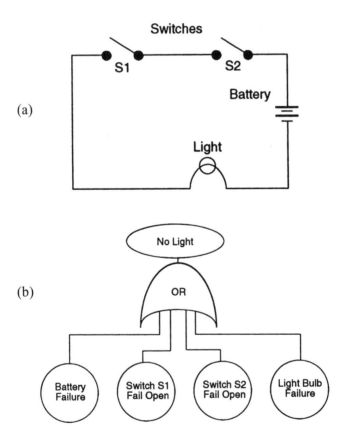

Fig. 5.2. Fault tree example of a simple light circuit. (a) Simple light circuit. (b) Fault tree for the light circuit. (From R&M Guideline)

The idea behind an FTA is to think about all possible ways in which an undesirable event could occur, and then ultimately find all means to prevent the failure from occurring. In some manner, the FTA is the reverse of an FMEA, which is a bottom-up approach. The FMEA starts with the origins and causes of a failure and then analyzes them to prevent them from occurring. In contrast, the FTA starts at the top and then considers the causes of failure. The FTA sometimes is referred to as a backward FMEA.

In effect, the FTA is an analytical tool that does the following:

1. Identifies and relates to the problem undesirable event that could result in

   - Loss of the system
   - Substantial damage to the system
   - A safety critical situation

2. Assesses the effects of the system safety or environmental changes

3. Detects the conditions that would cause a failure

4. Provides a math model probability of occurrence of an unwanted top event

5. Identifies potential safety problems

6. Communicates and supports

   - Tradeoffs
   - System design tolerable decisions

7. Recommends corresponding solutions to the potential problems

8. Is a deductive logic diagrammatic analysis on a system and/or a process failure

9. Is an alternative approach for determining system reliability, although analysis often can become complicated

The fault tree figures encompass logic systems derived from Boolean algebra and event representative symbols. Figure 5.3 shows the conventional symbols that are used.

FTA Symbol	Name	Causal Relationship or Description
	AND Gate	Output event occurs if all the input events occur simultaneously. Includes indicatives such as and, also, although, besides, both, etc.
	OR Gate	Output event occurs if any one of the input events occur. Includes indicatives such as or, at least one of, either, unless, etc.
	Top	Contains the description of the system-level fault or undesirable event.
	Intermediate fault	Contains brief description of a lower level fault or failure events combined through a logic gate.
	Input	Contains an input fault to the system; this input car be either a condition from a source outside the system or the normal failure of a system component, and is expected to occur.
	Basic	The lowest level of event under investigation or the primary failure event.
	Undeveloped	An event which we do not wish to expand further due to its insignificance or lack of empirical data.
	Not displayed	Event identified but not displayed.
	Similar transfer	A transferred event that is similar in function, but includes one or more events that are different from the second location.
	Logic transfer	An event that is transferred in from above or is transferred out when from the side.
	Same event	An identical event transferred from another location of the tree.

Fig. 5.3. Fault tree logic symbols.(From R&M Guideline)

There is nothing that states that the manufacturer must use these fault tree logic symbols in performing an FTA, but these are conventional and understood by most users. If the manufacturer wishes to use its own symbols, in the same way that the manufacturer can use its own flowcharts and other diagrams, it should be free to do so. However, others interpreting the charts also should understand how the figures are being used. In addition, the depth to which an FTA is performed depends on the requirements set by the customer. Most users probably will be satisfied with the pictorial representation of the FTA rather than a fully developed study. Figure 5.4 provides an example of a safety concern on a production machine.

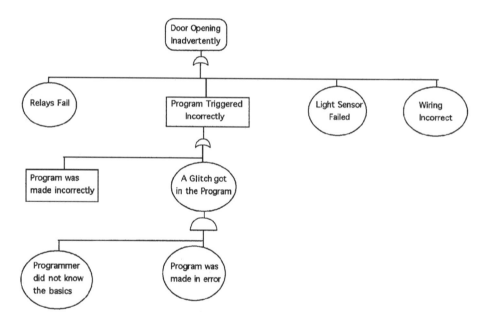

Fig. 5.4. Fault tree analysis (FTA) for a safety concern.

The comparisons shown in Table 5.1 further describe the differences between an FMEA and an FTA.

TABLE 5.1
COMPARISON OF DIFFERENCES BETWEEN AN FMEA AND AN FTA

FMEA	FTA
• Starts with the origin of a failure mode and then finds the causes, probabilities, and corrective actions	• Starts with an accident or an undesirable failure event, then determines the causes followed by the origins of those causes, and then determines what can be accomplished to prevent the failure
• Considers all potential failure modes that can be produced through experience and analysis	• Studies only negative outcomes to warrant further analysis
• Uses an inductive approach	• Uses a deductive approach
• Requires less engineering	• Requires skilled personnel
• Includes a limited safety assessment	• Manages risk assessment and safety concerns
• Does not delineate failure paths	• Provides a good assessment of the failure paths, with control points well enhanced
• Looks at each failure mode separately	• Demonstrates a more selective method of showing the relationship among the events that interact with each other

The Rome Air Development Center (RADC) of Rome, New York, has formulated a seven-step approach to generate an FTA. These steps are shown here to assist manufacturers in developing their own plans.

1. Define the system, ground rules, and any assumptions to be used in the analysis.

2. Develop a simple block diagram of the system, showing inputs, outputs, and interfaces.

3. Define the top event (ultimate failure effect or undesirable event) of interest.

4. Construct the fault tree for the top event, using the rules of formal logic.

5. Analyze the completed fault tree.

6. Recommend any corrections for design changes.

7. Document the analysis and its results.

An FTA is one of the few tools that can depict the interaction of many factors, and it also manages to consider the events that would trigger the failure or undesirable event. Figure 5.4 showed how a safety concern of the machinery being analyzed was considered in the study. Designers should consider the safety aspects of their designs early in the concept stages to ensure that necessary changes can be accomplished before any undesirable events have a chance to occur. We have seen how FMEA and FTA differ, and where one may be more effective than the other. The user typically defines what method of analysis is required, whereas the manufacturer follows the specified directions.

In conclusion, an FTA can be a useful tool for identifying conditions that can lead to a component safety problem, a malfunction, or other undesirable event. Engineers should not think that an FTA will solve all their problems. However, the process forces the designer to review the possibilities of the event failures.

References

Cohen, Daniel I.A., *Introduction to Computer Theory*, John Wiley and Sons, New York, 3rd ed., 1997.

Juran, Joseph M., *Juran's Quality Handbook*, 5th ed., McGraw-Hill, New York, 1999.

Nelson, R.J., *Introduction to Automata*, John Wiley and Sons, New York, 1967.

Chapter 6

Quality Function Deployment

Quality Function Deployment (QFD) is a discipline for product planning and development or for redesigning an existing product in which key user wants and needs are deployed throughout an organization. It can be thought of as a conceptual map that provides the means for cross-functional planning and communication, and a method of transforming customer wants and needs into quantitative terms. The discipline should be as a planning process, not a tool for solving problems (although in unique situations it may be).

Quality Function Deployment provides a structure for ensuring that users' wants and needs are carefully heard and then directly translated into the internal technical requirements of a company—from component design through final assembly. The minimum objective of a QFD exercise is to satisfy the needs of the customer; however, manufacturers should think beyond this minimum and create quality that is beyond the competition while remaining within rival pricing. Some quality professionals refer to QFD as the "House of Quality." Figure 6.1 shows a pictorial view of the makeup of a typical QFD layout.

The Japanese originated QFD in 1972 at the Mitsubishi Kobe Shipyard. The methodology received further development by Toyota and one of its suppliers. It then was used successfully by other Japanese manufacturers. Toyota gained significant cost savings and quality improvements using the technique. More than a decade later, U.S. companies realized the benefits of the procedure and were able to use it to their advantage also. Manufacturers soon learned that QFD was a meaningful method that saved money while simultaneously conforming to customer requirements. Manufacturers began to listen to the customer.

In effect, QFD brings the needs and desires of the customer into focus at the initial stage of product development. It does not focus on what is wrong with an existing product; rather, it focuses on what the customer wants or expects. It brings the voice of the customer into the process by coupling the manufacturer to

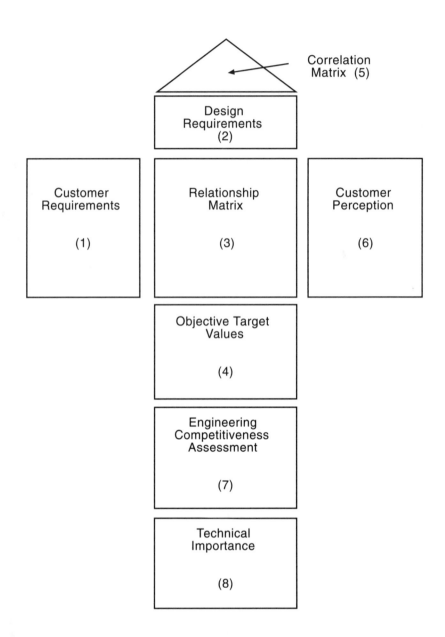

Fig. 6.1. Quality Function Deployment (QFD) conceptual map, also known as the "House of Quality."

the user. Marketing staff, designers, manufacturing staff, and the customer, if it can be arranged, all work closely together from the time the product is first conceived to when it is in the hands of the ultimate user. If utilized properly, QFD more or less forces an organization to turn its attention toward the wants of the customer rather than what the organization thinks is required.

The theory behind QFD is that the design and manufacturing engineers listen to the voice of the customer. The technique encompasses a series of interlocking matrices that translate customer concerns and desires into a product or process characteristic. When appropriate additional information is added to the matrix— such as correlation between parameters, units in some form of priority rating, and competitive appraisals—QFD provides the means in which a cross-functional team can plan and communicate to the benefit of the customer. It can detect and solve potential problems before those problems occur.

When QFD is developed, it is a team effort. There is nothing magical about the House of Quality, although it appears somewhat complicated. Some effort is required in its development and in the coordination of team members. After its construction, the layout is similar to a road map that contains the customer's attributes, the engineering features to support those attributes, the processing necessary to manufacture the features, and the production requirements. The overall picture of the house can be viewed together with the details that compose it.

The study of the House of Quality provided some clues as to how the "house" should be constructed. First, the voice of the customer is noted. This can be in the form of needs or what the customer believes are the requirements. These are generated into product features and then into process characteristics. Next, the manufacturing and quality engineers must consider how the processes can be managed to ensure that the voice of the customer has its continuity from initial design to the production of the product.

A properly developed QFD ensures that the customer needs translate into both the design of the product and the design of the process. In other words, the product is manufactured per the design rather than one that has nonconforming characteristics. Design and product specifications assure a customer of the desired item, but the item must be manufactured suitably. It is also reasonable to assume that QFD contributes to a robust design when using this approach.

Some QFD scenarios can develop into massive, complex "houses." Some requirements can be conflicting and time consuming. The many concerns of the customer should all be taken into account and each of the parameters duly considered as part of the design. Then the processes must be adequately designed to ensure that the manufacturing process guarantees the design. In addition, QFD compares the design features against those of the competition, allowing for the potential reduction of lead times. Questions at the start of a QFD generally include the following:

1. What do customers want (attributes)?

2. What are the preferences (priorities)?

3. Will responding to the needs yield a competitive advantage (value)?

4. What characteristics match customer attributes?

5. How does each characteristic affect each customer attribute?

6. How does one change affect the other characteristics?

The following steps are necessary to construct the House of Quality (the steps follow the numbering as in Fig. 6.1).

1. **Customer Requirements:** The customer's needs and concerns first are noted, indicating their relative importance. Customer perception is shown for each customer concern. The manufacturer will have to determine how to respond to each voice of the customer. Acquiring the voice may require some ingenuity on the part of the manufacturer, but it should not be a difficult task if planned thoroughly. The manufacturer can have a cross-functional team that concentrates on what the team believes are the customer's needs. Some team members can visit with the customer. Interviews, questionnaires, product training and educational seminars, personal observations, and society meeting and trade show exchanges are examples of how manufacturers can have interchanges that seek to determine the basic needs and wants of their customers. Figure 6.2 shows the identification of the customer necessities and concerns.

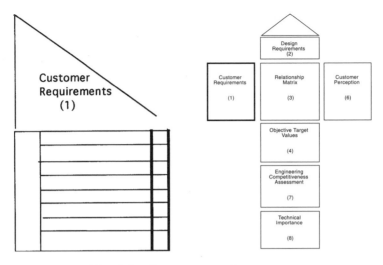

Fig. 6.2. Customer requirements.

2. **Design Requirements**: The engineering features are noted. This is accomplished by developing the technical information by matching it with, and translating it to, the customer portion of the matrix. It is important to respond to each voice of the customer. These should be measurable features and include design and quality requirements, product attributes or criteria, and technical requirements. In turn, these are examined to determine how the technical requirements may affect customer desires. Figure 6.3 depicts the engineering features.

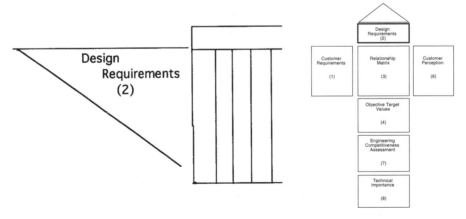

Fig. 6.3. Engineering features.

71

3. **Relationship Matrix:** Interrelationships among the customer concerns and engineering features are noted with symbols indicating strong positive to strong negative connections. This relationship is depicted where the customer and technical portions intersect with one another. The cross-matching provides an opportunity to record the existence and strengths in the relationships. The symbols that indicate the varying strengths are shown below.

Symbols:

⊙ = Strong positive relationship

○ = Moderate relationship

Δ = Weak relationship

X = Strong negative relationship

Figure 6.4 shows the dependence among the customer concerns and engineering features.

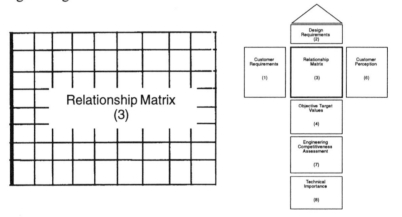

Fig. 6.4. Relationship among customer concerns and engineering features.

4. **Objective Target Values:** Objective measures are added below the relationship matrix. These are coordinated with the engineering features and show the technical assessments units or how the manufacturer ensues against its competition. Figure 6.5 depicts the objective measures.

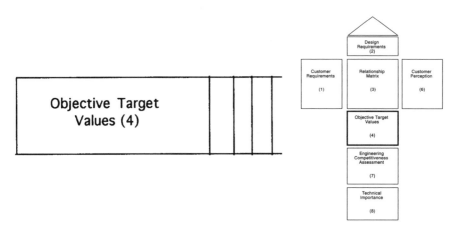

Fig. 6.5. Competitive objective assessment.

5. **Correlation Matrix:** The "roof of the house" is added to the structure, coordinating the engineering features by indicating the relationship with the checkmarks. These can be thought of as tradeoffs by comparing the technical requirements against each other and capturing the interactions among the design requirements. These are examined to determine the projected result of changing one requirement against another and also to manage conflicting requirements. Figure 6.6 portrays the coordinated engineering features.

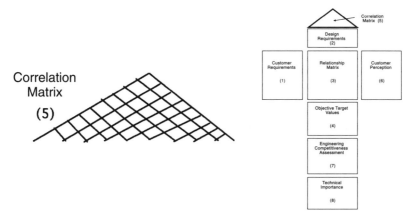

Fig. 6.6. Coordination of the engineering features.

73

6. **Customer Perception:** This block depicts the customer concerns and cognizance of the manufacturer. It also can include customer complaints and information from competitors, which may prove useful. Figure 6.7 shows the user concerns and problems that the user may have had with past designs.

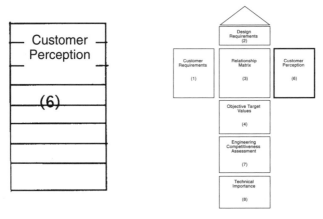

Fig. 6.7. Customer perception.

7. **Engineering Competitiveness Assessment:** Below the objective target values is the assessment of the engineering features that relate to the characteristics offered by the manufacturer's competitors. This provides some measure of how the manufacturer's design relates to those of its rivals. Figure 6.8 shows the assessment chart.

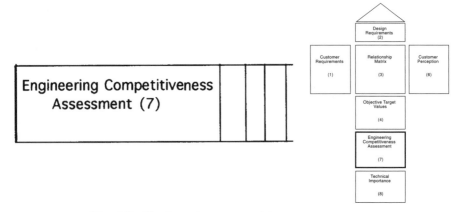

Fig. 6.8. Engineering competitiveness assessment.

8. **Technical Importance:** The basement of the house scales the importance of the design features. These are shown in absolute and relative (percentages) importance or priority value, and are some measure of which characteristics should be targeted. This section provides a weighing or prioritizing when reviewing the overall relationships in the house above this basement. Figure 6.9 represents the foundation of the House of Quality, with the added technical importance of the design features.

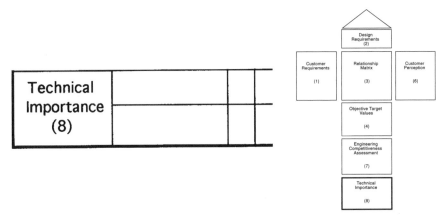

Fig. 6.9. QFD importance of the design features.

When the House of Quality is completed, the cross-functional team of marketing, quality and design engineering, and manufacturing engineering can probe deeply into the analysis. The focus of attention started with the customer requirements and continued through the final implementation of the processes in the shop. The idea is to satisfy the needs of the customer. The team can view the whole process as a conceptual map that presents an overall picture while simultaneously showing the fine details. In some sense, the complete QFD diagram is similar to a road map. By now, process manufacturing should have been well finalized as it was developed concurrently with the product design.

Looking at the overall picture, we can see how QFD shows as a real benefit to the manufacturer and how the customer needs are followed. It gives the manufacturer the opportunity to be proactive rather than reactive. Fewer design changes will occur in the production process because much thought was put into the design early in the process, as well as the manner in which it coordinated with manufacturing. Equipment tooling manufacturers should keep in mind that although they initially meet the needs of their customers, they also must be continuously alert and responsive to changes that users may want. Levels of satisfaction should be updated as much as possible. The end result is higher customer satisfaction, lower cost due to fewer failures, and an increase in employee confidence.

Other benefits and advantages can be realized when manufacturers use QFD as part of their planning and development. We discussed in depth the understanding of the customer and how costs would be lowered. One Japanese company showed how the use of QFD narrowed the necessity of design changes to the first few months of development rather than stretched to the point when production started and even beyond that step. Reduced lead time and minimum launch errors generally occur when incorporating QFD. Communication and cooperation among various disciplines improve. As a result of enhanced quality, warranty costs are reduced substantially, with corresponding increases in market share. All these benefits and more materialize when QFD is employed with commitment.

One of the concerns of a manufacturer may be that a QFD exercise is too complicated and costly for its machinery. There is no question that QFD will involve several key players and will take time. However, the end result will be a cost savings and a more satisfied customer. By determining the user requirements early in the process, the manufacturer is more able to address the user's concerns early in the concept phase. What was proposed in this chapter is the classic QFD method, which does not imply that all QFDs must be accomplished in the same manner. Manufacturers can be innovative and develop their own designs that may be more appropriate for their products.

Quality Function Deployment was discussed more in the vein of what engineering can do with the customer desires. It also can be used for a business plan, improvement in production methods, proposal systems, and where a problem keeps occurring, although QFD should not be thought of as a problem-solving tool. It is more reasonable to think of QFD as being a disciplined approach to solving potential quality problems before the design phase of the machinery. One of the advantages of QFD not mentioned previously is that manufacturers are discovering what the customer says he or she desires is limited by the customer's existing knowledge. Quality Function Deployment goes beyond the customer's awareness, which gives the manufacturer an opportunity to develop new products. These may relate more to the customer's potential needs. Many new and exciting ideas have developed in this manner. Quality Function Deployment gives an organization an opportunity to do this and much more.

References

Day, Ronald G., *Quality Function Deployment—Linking the Company with Its Customers*, ASQC Quality Press, Milwaukee, WS, 1993.

King, Bob, B*etter Design in Half the Time: Implementing QFD in America*, GOAL/OPC, Metheum, MA, 1987.

Chapter 7

Reliability Testing

The ultimate proof of any product is its ability to perform to its intended functional specification to the satisfaction of the customer. The broad terminology of reliability testing encompasses three main forms of testing:

1. Development
2. Qualification
3. Acceptance

However, it also extends to include prototype, performance, environmental factors, stress, demonstrated life, accelerated life, and production. The customer is the ultimate tester, although most organizations do not think of the customer as the person who performs a reliability test. All testing relates to reliability and maintainability in one form or another.

An explanation of the three main groupings of reliability testing is provided in the following sections. The primary purpose of a development test is to ascertain the feasibility of a specific design. It is accomplished mostly for obtaining information to establish the design criteria. This provides the initial information required in the establishment of the components that make up the design. Testing usually is accomplished by the machine builder and should be at least as rigorous as what the tooling and equipment (TE) user would impose. All environmental features, loads, cycles, stresses, feeds, and similar factors should be part of the development test.

The qualification test provides data to determine the capability of the production process. This does not necessarily mean the machinery is made on the "production line," but it should be made with "production tooling." The idea is to further check the design and also to examine the manufacturing capability of the product. The test infers the existence of acceptance criteria that the product is expected to meet. Care must be exercised to ensure that the production tooling is as close to the production line processing as possible. If it is not, the manufacturer is kidding itself to think that production workers will be as meticulous as experienced, well-trained, and highly paid machinists, technicians, inspectors, and assemblers. The management of who conducts the qualification test usually is determined by mutual agreement between the TE user and the manufacturer. Qualification testing sometimes can be a misnomer because production tooling may be used to fabricate and assemble the product, but it may not represent a "production build line."

Acceptance testing is the formal test that determines whether the final product meets all criteria for which it is designed. In some cases, it is the testing that is required by the customer on the manufactured machinery from the production line. It also is the testing required by the customer, in which all design and R&M functional specifications must meet product requirements before the customer will purchase the machinery. Acceptance testing usually is conducted by the customer, but it can be contracted to be performed by the manufacturer or an agreed upon outside test facility. In any event, the test results should be monitored closely by both the TE user and the manufacturer.

The *R&M Guideline* defines an acceptance test (qualification test) as "a test to determine machinery/equipment conformance to the qualification requirements in its equipment specifications." Later the *Guideline* separates the two test methods in the Recommended R&M Training Matrix in Appendix B of the book. Many manufacturers use the two terms interchangeably. However, the difference in nomenclature becomes much clearer if you think of qualification testing as tests done on prototypes or early development models (i.e., models that are made with production tooling), and acceptance testing as tests done on early production units.

The idea of reliability testing is twofold:

1. To validate the life of the machinery
2. To determine if there are failure modes in the system

To accomplish these requirements, the design engineer must plan with the test engineers, customer, and other concerned personnel. Many factors must be considered in this planning, such as what to look for, what tests are to be performed, what environments are to be used, and what sample sizes would be representative. The design engineer should have a good idea of how he or she is planning to interpret the data when information initiates from the testing.

Testing provides some measure of evaluating the design at any stage of development. Environmental tests evaluate the ability of the product to withstand the expected condition in which the product will be used. Stress tests determine the stress level the product can withstand. In the concept of Environmental Stress Screening (ESS), planned overstress conditions are applied to determine the influence of overstressing and also for accelerating the test. When using accelerated tests, the design engineer should have a good measure of the correlation between overstressing and the actual life of the product. One problem with accelerated tests is that no correlation usually exists between the test results and the results of actual performance. Nonetheless, it is utilized often because it usually is less costly.

MIL-STD-785B, Requirements for Reliability Program (Systems and Equipment), also includes ESS as one of the four areas of testing. For some military contracts, ESS requires tests that are conducted above or below the stated environmental conditions, in an attempt to identify early failures. Commercial businesses have accepted the concept of ESS, which is also referred to as "shake and bake" testing. To more fully ensure the meeting of reliability requirements, some customers may require 100% ESS be performed on critical parts, assemblies, and equipment. Testing is usually nondestructive but will detect the weak item and manufacturing defects, and eliminate early field failures.

The idea of testing is to identify potential reliability problems early in development and reduce field failures. Environmental Stress Screening can include tests such as high and low temperatures, shock and vibration, increased and reduced pressures, electrical variability jolts, salt fog, fungus growth, chemical attack, solar radiation, rain and hail, humidity, and sand and dust. The customer usually determines what testing is to be done, but manufacturers also can do the same if they wish to know how their products perform outside specification conditions.

Closely related to ESS is accelerated life testing. The technique often is viewed as a relatively new procedure, but it has been in use for some time. The concept of accelerated life testing initiated during World War II, when the military required a rapid acceptance technique to validate equipment that was to be used in all locations of the world and under extreme environmental situations. Later, in the 1970s, commercial organizations saw the potential of the technique and incorporated it into their test programs. Today, accelerated life testing is used universally.

In effect, accelerated life testing provides a quick way to evaluate a design. By overstressing specific components through extreme temperature cycles, high physical and electrical loadings, software applications, and other overexertions, the life of components can be estimated under the overstressed conditions. In addition, the testing assists in predicting the projected life of the machinery and how maintenance actions prolong the life of the product. Accelerated testing and ESS both help in finding the main failure modes. However good the technique, the engineer should use discretion in estimating the total life of the machinery.

How the manufacturer performs its reliability testing is determined largely by a combination of its own planning and what the customer would like to see. As discussed previously, straightforward life testing to customer requirements can be accomplished. Accelerated life testing and ESS also can be performed. However, everyone should understand that these tests usually are performed within overstressed conditions. Another similar form of accelerated life testing is the acquisition of field loads via stress collection devices, putting this information into an automatic format that can operate the machinery for 24 hours per day under the same or similar conditions in the laboratory. Some manufacturers call this bench testing. The testing is not accelerated *per se*, but it is continuous with the loads and situations that the machinery will experience in actual use. This type of testing provides real-time performance evaluation and, if accomplished properly, is reflective of actual user conditions.

If the reliability predictions, design reviews, FMEAs, and other analytical methods were accomplished in depth and with fair accuracy early in the design phase, testing proceeds smoothly in most cases. If the initial analytical studies were incomplete, you can expect only problems during the testing. The concept of Analyze-Test-Analyze-And-Fix (ATAAF) invariably results in a superior quality product than does the concept of Test-Analyze-And-Fix (TAAF) and generally is less expensive in the long run.

Maintainability tests determine the time required to make repairs (Mean Time To Repair, or MTTR) and provide for Maintenance Ratios (MR) when desired. (The MR usually is the ratio of repair time to total operating time.) Tests that are run to failure generally determine the wear-out time for a product and the failure modes associated with the test run.

The QS-9000TE machinery suppliers may be required to perform special tests for the purpose of verifying the R&M requirements specified in their procurement documents. Mutual agreement between the customer and supplier should determine the test conditions, such as duration, operating conditions, and any extra parameters that are to be included. Test programs usually extend to approximately four times the specified Mean Time Between Failures (MTBF), with all failures documented and maintenance times recorded. All reliability testing should be detailed in the R&M plans.

The *R&M Guideline* recommends that the supplier should perform a preliminary assessment of R&M before the machinery is ready to begin acceptance testing. The supplier should verify that all important machine parameters are operating within their specification limits. If any infant mortality-type problems exist, they should be uncovered during this testing and corrected before proceeding further.

Chapter 2 presented several distribution curves that can be used for reliability testing. One of the most prevalent and beneficial distributions that is utilized is the Weibull distribution which can closely simulate other distributions. Weibull plotting is a fairly simple technique for analyzing failure data, and it allows an extrapolated process for these failures that occurred under testing.

The Weibull analysis technique is presented in the latter part of this chapter. Weibull analysis is an extremely versatile and useful distribution that has proven itself valuable in reliability applications. The analysis results in the derivation of three parameters:

1. Scale
2. Slope
3. Location

The scale parameter is the characteristic life—the average life, Mean Time To Failures (MTTF), or other measure of life—minus the location or minimum life. The slope defines the shape parameter of the distribution of data points in the Weibull graph. It also helps describe the observed variability. The location or minimum life is the initial period of no failures. In most cases, it is considered zero. The instructions for plotting and analyzing the Weibull graph and parameters are further explained at the end of this chapter.

Some risks are involved with testing, of which TE machine suppliers and customers should be aware. One consideration is whether the equipment will be utilized as originally intended or will it be overloaded or misused? Another problem, which is true not only for machine builders but for many products, is the meticulous fabrication, assembly, and testing of early design models that do a tremendous performance job. Later, when production models are made, they "fall flat on their faces." Does the customer and supplier understand this relationship? The extra care that is exercised during the prototype build generally is not practiced on the production line as it is with the technicians and more qualified assemblers who must be very careful about following the instructions outlined by the manufacturing engineers. This is another reason for conducting the qualification and acceptance tests. Regardless, both manufacturers and customers should be aware of the potential inconsistencies.

One serious consideration for designers is not to depend on a small number of tests to substantiate large production runs. The evaluation of test results is a key parameter to determining the real reliability of a product. Often, because of the pressure of delivery, shortcuts are taken in testing and in analyzing data. All failure data should be analyzed and corrective action should be implemented before initial release of the machinery or tools. Remember that the later potential problems are detected, the more costly the corrections are.

As part of the overall reliability testing program, detailed planning must occur. This is true for reliability testing as well as for quality and reliability strategic planning. The key characteristics must be identified. The level of testing, environmental conditions, sample sizes, statistical aspects, and other factors all must be part of the planning. Both the customer and the manufacturer usually are involved in the planning. The end result should provide a good idea of what must be accomplished so that the product reliability meets or exceeds the required value.

Instructions for Plotting and Analyzing Failure Data on a Weibull Probability Chart

The Weibull analysis technique is useful for examining test data and graphically displaying the information on Weibull probability paper. The technique provides a means of estimating the percent failed at specific life characteristics, together with the shape of the failure distribution. The following procedure presents a manual method of utilizing the analysis; however, many computer programs can do the same calculations and also plot the Weibull curve. Weibull analysis is one of the simpler analytical methods, but it is also one of the most beneficial.

Weibull analysis can be used for applications other than simply analyzing failure data. It can be used for comparing two or more sets of data, such as different designs, materials, or processes. The following are instructions for conducting a Weibull analysis.

1. Gather the failure data (e.g., in miles, hours, cycles, number of parts produced on a machine), and then list in ascending order.

2. Using the Table of Median Ranks (shown at the end of this chapter), find the column corresponding to the number of failures in the sample tested. In our example, we have a sample size of ten which utilizes the "sample size 10" column. The "% Median Ranks" then are read directly from the table.

3. Match the hours (or some other failure characteristic that is measured) with the median ranks from the sample size selected. For example,

Actual Hours to Failure	% Median Ranks	
95	6.7	
110	16.2	
140	25.9	Sample size
165	35.5	of 10 failures
190	45.2	
205	54.8	
215	64.5	
265	74.1	
275	83.8	
330	93.3	

4. In constructing the Weibull plot, label the "Life" on the horizontal log scale on the Weibull graph in the units in which the data were measured. Try to center the life data close to the center of the horizontal scale (see Fig. 7.1).

5. Plot each pair of "Actual Hours to Failure" (on the horizontal logarithmic scale) and "% Median Rank" (on the vertical axis, which is a log-log scale) on the graph. The matching points are shown as dots ("•s") on Fig. 7.1. Draw a "line of best fit" (generally a straight line) as close to the data pairs as possible. Half of the data points should be on one side of the line; the other half of the data points should be on the other side of the line. No two people will generate the exact same line, but analysts should keep in mind that this is an "eyeballed" estimate. Design engineers also should keep this in mind when quoting reliability parameters. If you wish to obtain more precision (but remember that you are still working with sample sizes), a computer program should be used or manual calculations should be done.

6. After the line of best fit is drawn, the life at specific points can be found by going vertically to the "Weibull line" and then going horizontally to the "Cumulative % Failed." In other words, this is the percent that is expected to fail at the life that was selected. In Fig. 7.1, 100 was selected as the life. Then, going upward to the line and then across, we can see the expected percent failed to be 10%. In this case, the life at 100 hours is also known as the B_{10} life (or 90% reliability) and is the value at which we would expect 10% of the parts to fail when tested under similar conditions.

7. The Weibull graph can be used for estimating the cumulative percent failure at a specified life, or it can be used for determining the estimated life at a cumulative percent failure. In Fig. 7.1, we would expect 63.2% of the test units to fail at 222 hours. This value at 63.2% also is known as the Characteristic Life or the Mean Time Between Failures (MTBF) for the example distribution. Or, looking at the chart another way, we would like to estimate the failure hours at a specified percent failure. For example, at 95% cumulative percent failed, the hours to failure is 325 hours. After the Weibull plot is determined, an analyst can proceed either way.

8. The Weibull graph also can be used to estimate the reliability at a given life. If the designer wishes to estimate the reliability of life at 200 hours,

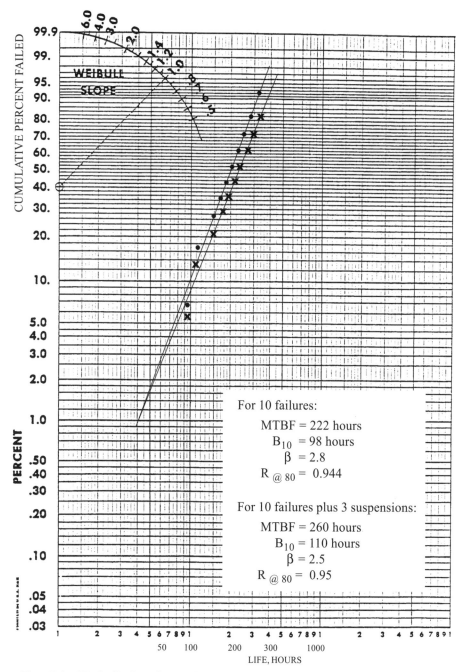

For 10 failures:

MTBF = 222 hours
B_{10} = 98 hours
β = 2.8
$R_{@\,80}$ = 0.944

For 10 failures plus 3 suspensions:

MTBF = 260 hours
B_{10} = 110 hours
β = 2.5
$R_{@\,80}$ = 0.95

Fig. 7.1. Weibull plots for 10 failures and for 10 failures plus 3 suspensions.

he would go vertically to the Weibull line and then go horizontally to 52%, which is the percent expected to fail. The estimated reliability at 200 hours would be

$$1 - 0.52 = 0.48$$

or 48%. At 80 hours, it would be

$$1 - 0.056 = 0.944$$

or 94.4%. The slope (β) is obtained by drawing a line parallel to the Weibull line on the Weibull slope scale that is in the upper left corner of the chart. These parameters plus the B_{10} life can be shown on the chart. The reliability also can be calculated using the Weibull distribution equation (see Chapter 2 for explanation)

$$R = e^{-\left(\frac{80}{222}\right)^{2.8}} = 0.9442$$

9. If a computer program is used, the computer determines the calculation for the line of best fit. Some programs draw the graph and show the paired points, the line of best fit (using the least-squares method or the maximum-likelihood method), the reliability at a specified hour (or other designated parameter), and the slope of the line.

10. One interesting observation regarding the Weibull graph is the interpretations that can be made about the distribution by the portrayal of the slope. When the slope is:

- Less than 1, this indicates a decreasing failure rate, early life, or infant mortality.

- Approximately 1, the distribution indicates a nearly constant failure rate (useful life or a multitude of random failures).

- Exactly 1, the distribution has an exponential pattern.

• Greater than 1, the start of wear-out is evident.

• Approximately 3.55, a normal distribution pattern is found.

11. Weibull plots can be made if test data also include test samples that have not failed. Parts that have not failed (for whatever reason during the testing) can be included in the calculations, together with the failed parts or assemblies. The nonfailed data are referred to as suspended items. The method of determining the Weibull plot is shown in the next set of instructions.

Instructions for Plotting Failure and Suspended Items Data on a Weibull Probability Chart

1. Gather the failure and suspended items data, include the suspended items, and list in ascending order.

Item	Hours to Failure or Suspension	Failure or Suspension Code*	
1	95	F1	
2	110	F2	
3	140	F3	
4	165	F4	Sample Size of 13:
5	185	S1	10 failures,
6	190	F5	3 suspensions
7	205	F6	
8	210	S2	
9	215	F7	
10	265	F8	
11	275	F9	
12	330	F10	
13	350	S3	

* Code items as failed (F) or suspended (S).

2. Calculate the mean order number of each failed unit. The mean order numbers before the first suspended item are the respective item numbers in

the order of occurrence (i.e., 1, 2, 3, and 4). The mean order numbers after the suspended items are calculated by the following equations:

Mean order number = (Previous mean order number) + (New number)

where

$$\text{New increment} = \frac{(N+1)-(\text{Previous mean order number})}{1+(\text{Number of items beyond present suspended item})}$$

and

$$N = \text{total sample size}$$

For example, to compensate for S1 (the first suspended item),

$$\text{New increment} = \frac{(13+1)-4}{1+8} = 1.111$$

and the mean order number of F5 (the fifth failed item) is

$$4+1.111=5.111$$

Note that only one new increment is found each time a suspended item is encountered.

The mean order number of F6 (the sixth failed item) is

$$5.111+1.111=6.222$$

The new increment for the mean order number of F7 (the seventh failed item) is

$$\frac{(13+1)-6.222}{1+5}=1.296$$

Then, the mean order number of F7 (the seventh failed item) is

$$6.222+1.296=7.518$$

and so forth for F8, F9, and F10.

This new increment also applies to mean order numbers, as shown here.

Item	Hours to Failure or Suspension	Failure or Suspension Code*	Mean Order Number
1	95	F1	1
2	110	F2	2
3	140	F3	3
4	165	F4	4
5	185	S1	—
6	190	F5	5.111
7	205	F6	6.222
8	210	S2	—
9	215	F7	7.518
10	265	F8	8.815
11	275	F9	10.111
12	330	F10	11.407
13	350	S3	—

* Code items as failed (F) or suspended (S).

3. A rough check on your calculations can be made by adding the last increment to the final mean order number. If the value is close to the total sample size, the numbers are correct. In our example,

$$11.407 + 1.296 = 12.702$$

which is a close approximation to the sample size of 13.

4. Using the Table of Median Ranks for a sample size of 13, we can determine the median rank for the first four failures (shown on page 96 in this chapter), or we can use the approximate median rank formula.

$$\text{Median rank} = \frac{J - 0.3}{N + 0.4}$$

where

$$J = \text{Mean order number}$$
$$N = \text{Total sample size}$$

For example, the median rank of F5 is

$$\frac{5.111 - 0.3}{13 + 0.4} = 0.359$$

and the remainder of the failures is

$$\frac{6.222 - 0.3}{13 + 0.4} = 0.442$$

$$\frac{7.518 - 0.3}{13 + 0.4} = 0.539$$

and so forth.

Item	Hours to Failure or Suspension	Failure or Suspension Code*	Mean Order Number	% Median Rank
1	95	F1	1	5.2
2	110	F2	2	12.6
3	140	F3	3	20.0
4	165	F4	4	27.5
5	185	S1	—	—
6	190	F5	5.111	35.9
7	205	F6	6.222	44.2
8	210	S2	—	—
9	215	F7	7.518	53.9
10	265	F8	8.815	63.5
11	275	F9	10.111	73.2
12	330	F10	11.407	82.9
13	350	S3	—	—

* Code items as failed (F) or suspended (S).

5. Label the "Life" on the horizontal log scale on the Weibull graph in the units in which the data were measured. Try to center the life data close to the center of the horizontal scale (see Fig. 7.1).

6. Plot each pair of "actual hours to failure" (on the horizontal scale) and "% median rank" (on the vertical scale) on the graph. These points are shown as Xs on Fig. 7.1. Draw a line of best fit (generally a straight line) as close to the data pair as possible. Half of the data points should be on one side of the line, and the other half of the data points should be on the other side of the line.

7. After the line is drawn, the life at specific points can be found by going vertically to the Weibull line and then going horizontally to the "Cumulative % Failed." In other words, this value is the percent that is expected to fail at the life that was selected. In the example, 200 hours was selected as the life. Then going upward to the line and then across, we can see the expected percent failed to be 40%.

8. As with the first set of data, other reliability parameters can be read
    directly from the Weibull plot:

$$\text{MTBF} = 260 \text{ hours}$$

$$B_{10} = 110 \text{ hours}$$

$$\beta = 2.5$$

Reliability @ 80 hours is $1 - 0.05 = 0.95$ reading from the graph,
or using the Weibull equation

$$R = e^{-\left(\frac{t}{\theta}\right)^{\beta}} = e^{-\left(\frac{80}{260}\right)^{2.5}} = 0.9488$$

9. Comparing the two examples on Fig. 7.1 shows that the analysis with sus-
    pended items results in a slightly higher reliability at 80 hours (0.94 vs. 0.95)
    and the characteristic life is larger (222 vs. 260). This uses the same failure
    data plus the three suspended items. The difference between reading from
    the chart and calculating from the Weibull equation indicates that these
    values are really estimates. Design engineers should keep this in mind
    and not be mislead into thinking that four-digit numbers are precise. More
    sophisticated methods can add to the analysis with confidence limits; these
    can be found in some of the references.

Additional Notes on the Use of the Weibull Analysis

1. Although Weibull plotting is an invaluable tool for analyzing life data,
    there are some precautions. Goodness-of-fit is one concern. This can be
    tested with various tests such as the Kolmogorov-Smirnov or Chi-square.
    The use of an adequate sample size is another concern. Generally, a sample
    size should be greater than ten; however, if the failure data are in a tight
    pattern (with relatively low variability), this regard can be a little more
    relaxing. Be suspicious of a curved line that best fits the data, as explained
    in Item 2 here.

2. After the Weibull plot is made and a curvilinear relation develops for the connecting points, it usually indicates that two or more distributions are making up the data. This may be due to infant mortality-type failures being mixed with the data, components from two different machines or assembly operations, or some other underlying cause. If a curved relationship is indicated, the analyst should revisit the data and try to determine if the data are made up of two or more distributions and then manage each distribution separately.

3. There is another parameter in the Weibull analysis that was not discussed. In addition to the shape or slope (β) of the Weibull line and the scale or characteristic life (the mean life or MTBF at the 63.2% cumulative percentage), there is the "location parameter." In most cases, it usually is zero and should be of little concern. In effect, it states that the distribution of failure times starts at zero time, which is more often the case because it is difficult to imagine otherwise.

4. One advantage of using Weibull analysis is that it is flexible in its interpretations. A wealth of information can be derived from it. If the Weibull slope is equal to one, the distribution is the same as the exponential, or a constant failure rate. If the slope is in the vicinity of 3.5, it is a "near normal distribution." If the slope is greater than one, the plot starts to represent a wear-out distribution, or an increasing hazard rate. A slope less than one generally indicates a decreasing hazard rate, or an infant mortality-type distribution.

5. Analysts should be careful about extrapolating beyond the data when making predictions. Remember that the failure points fall within certain bounds, and thus the analyst should have a valid reason when venturing beyond these bounds. When making projections beyond these confines, sound engineering judgment, statistical theory, and experience all should be deliberately planned.

Median Ranks

Rank Order	Sample Size									
	1	2	3	4	5	6	7	8	9	10
1	50.0	29.3	20.6	15.9	12.9	10.9	9.4	8.3	7.4	6.7
2		70.7	50.0	38.6	31.4	26.4	22.8	20.1	18.0	16.2
3			79.4	61.4	50.0	42.1	36.4	32.1	28.6	25.9
4				84.1	68.6	57.9	50.0	44.0	39.3	35.5
5					87.1	73.9	63.6	56.0	50.0	45.2
6						89.1	77.2	67.9	60.7	54.8
7							90.6	79.9	71.4	64.5
8								91.7	82.0	74.1
9									92.6	83.8
10										93.3

Median Ranks (continued)

Rank Order	Sample Size									
	11	12	13	14	15	16	17	18	19	20
1	6.1	5.6	5.2	4.8	4.5	4.2	4.0	3.8	3.6	3.4
2	14.8	13.6	12.6	11.7	10.9	10.3	9.7	9.2	8.7	8.3
3	23.6	21.7	20.0	18.6	17.4	16.4	15.4	14.6	13.8	13.1
4	32.4	29.8	27.5	25.6	23.9	22.5	21.2	20.0	19.0	18.1
5	41.2	37.9	35.0	32.6	30.4	28.6	26.9	25.5	24.2	23.0
6	50.0	46.0	42.5	39.5	37.0	34.7	32.7	30.9	29.3	27.9
7	58.8	54.0	50.0	46.5	43.5	40.8	38.5	36.4	34.5	32.8
8	67.6	62.1	57.5	53.5	50.0	46.9	44.2	41.8	39.7	37.7
9	76.4	70.2	65.0	60.5	56.5	53.1	50.0	47.3	44.8	42.6
10	85.2	78.3	72.5	67.4	63.0	59.2	55.8	52.7	50.0	47.5
11	93.9	86.4	80.0	74.4	69.5	65.3	61.5	58.2	55.2	52.5
12		94.4	87.4	81.4	76.1	71.4	67.3	63.6	60.3	57.4
13			94.8	88.3	82.6	77.5	73.1	69.1	65.5	62.3
14				95.2	89.1	83.6	78.8	74.5	70.7	67.2
15					95.5	89.7	84.6	80.0	75.8	72.1
16						95.8	90.3	85.4	81.0	77.0
17							96.0	90.8	86.2	81.9
18								96.2	91.3	86.9
19									96.4	91.7
20										96.6

Median Ranks (continued)

Rank Order	Sample Size									
	21	22	23	24	25	26	27	28	29	30
1	3.2	3.1	3.0	2.8	2.7	2.6	2.5	2.4	2.4	2.3
2	7.9	7.5	7.0	6.9	6.6	6.4	6.1	5.9	5.7	5.5
3	12.5	12.0	11.5	11.0	10.6	10.2	9.8	9.4	9.1	8.8
4	17.2	16.4	15.7	15.1	14.5	13.9	13.4	13.0	12.5	12.1
5	21.9	20.9	20.0	19.2	18.4	17.7	17.1	16.5	15.9	15.4
6	26.6	25.4	24.3	23.3	22.4	21.5	20.7	20.0	19.3	18.7
7	31.3	29.9	28.6	27.4	26.3	25.3	24.4	23.5	22.7	22.0
8	35.9	34.3	32.9	31.5	30.3	29.1	28.1	27.1	26.1	25.3
9	40.6	38.8	37.1	35.6	34.2	32.9	31.7	30.6	29.6	28.6
10	45.3	43.3	41.4	39.7	38.2	36.7	35.4	34.1	33.0	31.9
11	50.0	47.8	45.7	43.8	42.1	40.5	39.0	37.7	36.4	35.2
12	54.7	52.2	50.0	47.9	46.1	44.3	42.7	41.2	39.8	38.5
13	59.4	56.7	54.3	52.1	50.0	48.1	46.3	44.7	43.2	41.8
14	64.1	61.2	58.6	56.2	53.9	51.9	50.0	48.2	46.6	45.1
15	68.7	65.7	62.9	60.3	57.9	55.7	53.7	51.8	50.0	48.4
16	73.4	70.1	67.1	64.4	61.8	59.5	57.3	55.3	53.4	51.6
17	78.1	74.6	71.4	68.5	65.8	63.3	61.0	58.8	56.8	54.9
18	82.8	79.1	75.7	72.6	69.7	67.1	64.6	62.4	60.2	58.2
19	87.5	83.6	80.0	76.7	73.7	70.9	68.3	65.9	63.6	61.5
20	92.1	88.0	84.3	80.8	77.6	74.7	71.9	69.4	67.0	64.8
21	96.8	92.5	88.5	84.9	81.6	78.5	75.6	72.9	70.4	68.1
22		96.9	92.8	89.0	85.5	82.3	79.3	76.5	73.9	71.4
23			97.0	93.1	89.4	86.1	82.9	80.0	77.3	74.7
24				97.2	93.4	89.8	86.6	83.5	80.7	78.0
25					97.3	93.6	90.2	87.0	84.1	81.3
26						97.4	93.9	90.6	87.5	84.6
27							97.5	94.1	90.9	87.9
28								97.6	94.3	91.2
29									97.6	94.5
30										97.8

References

Johnson, Leonard, *The Statistical Treatment of Fatigue Experiments*, Elsevier Publishing Co., Amsterdam, 1964.

Juran, Joseph M., *Juran's Quality Control Handbook*, 5th ed., McGraw-Hill, New York, 1999.

Lewis, E.E., *Introduction to Reliability Engineering*, 2nd ed., John Wiley and Sons, New York, 1994.

O'Connor, Patrick D.T., *Practical Reliability Engineering*, 2nd ed., John Wiley and Sons, New York, 1985.

Chapter 8

Data Collection System and Failure Reporting, Analysis, and Corrective Action System

One of the key requirements for reliability and maintenance (R&M) enhancement is a dependable and effective data collection system. Some organizations feel that data reporting and analysis is the most important part of an R&M program. Tracking and feedback of component failures allow design engineers to recognize both the strong and the weak areas of their designs. During early development, much of the reliability testing is managed by the design engineer. It is crucial that the design engineer has a formal and organized feedback system in place, which relates to all problems that occur on the product.

The data collection system must be effective and efficient because copious amounts of data usually are generated during testing. If an efficient system for classifying the information does not exist, the information eventually can turn into a mishmash of data that would be extremely difficult to analyze. Data must be compiled, analyzed, and organized to determine trends, identify critical areas, and resolve design problems long before production of the equipment starts.

Coupled with an effective data collection system is a failure analysis and corrective action system. Most organizations call this system a Failure Reporting, Analysis, and Corrective Action System (FRACAS). The totality of the two systems blends together into a mutually agreed upon scheme between the supplier and the customer. The supplier and customer together can determine what kind of testing should prevail and what data and parts should be returned to the designer. Chapter 9 on failure analysis discusses the control and analysis of nonconforming parts. How data are collected and the manner in which the data acquisition is conducted can be in any arrangement, as long as the information and failed parts can be given to the designer, followed by the appropriate failure analysis and corrective action implementation.

The format for the compilation of data collection usually is the manufacturer's preference; however, the user may have some say about the kind of reports he or she would like to see. If the data collection is the supplier's responsibility, the user should be allowed to retrieve, or at least view, reports of concern. This is true even if the data are proprietary. Reports should include sufficient information to ensure the following:

- Maintenance actions are identified

- Time required to perform the maintenance tasks is included

- Coverage of failure details, number of persons needed for repairs, and the required support and test equipment used

- How the failure was discovered

- How the failure was isolated

- Any other information the tester or incident reporter feels would be beneficial in the analysis

Individual incident reports are the backbone of the FRACAS. The reports represent what the testing entails and the verification of what is happening to the new design, or a manufacturing error that was corrected. Figure 8.1 shows a brief summary of reported incidents. Information such as class of failure, mileage or hours, the incident description, root cause, corrective action, and status are useful in determining the overall status of the development program. Machine suppliers should have something similar to support the status of their projects, in order to be able to manage all the failures and maintenance information generated.

The individual incident reports should contain enough information to enable analysts to determine the cause of the failures and to arrive at appropriate fixes. Figure 8.1 is a partial summary of the reported incidents, providing a brief description of the complete actions that were done. The status column more or less puts the finishing touch on each incident because it indicates whether the problem remains "open" or has been "closed." This permits personnel involved with each incident to see at a glance if the corrective action has been implemented and if it is working as intended.

No.	Class	Mileage	Incident	Root Cause	Corrective Action	Status
1	Minor	0.0	The left rear door of the vehicle was found to be out of adjustment.	The shims for correcting adjustment are not shown in the technical manual.	The instructions for adjusting the shims have been added to the technical manual.	Closed
2	Minor	401.0	Oil cooler fins (upper rear center edge) were found to be bent.	This type of incident is accidentally caused by personnel when they are working in the engine compartment.	A note has been added to the manual, warning maintenance personnel not to bend these fins when working under the hood. A decal noting that personnel should be careful when working in the fin area has been added to the cooler.	Closed
3	Minor	2552.1	The windshield washer hose clamps screw threads were found to be damaged.	Hose clamp screw threads were interfering with the metal hood stripping when the hood was closed.	Screws will be shortened to prevent interference. ER 2209.	Open
4	Major	4730.0	The drive of the starter motor was found to be packed inside with mud.	Mud in the starter housing did not allow pinion engagement with the flywheel ring gear.	To prevent the entry of mud into the starter housing, a sealed torque cover has been released. EN3602.	Open
5	Major	5355.0	The fuel tank's drain plug was found to have a Class III leak.	The plug had a 1/4-inch section worn off. Suspect the incident was due to the plug extending too far out of the tank.	A newly designed recession is now built into the tank. ER5783.	Closed
6	Major	16361.0	The left rear lower ball joint was worn by dirt intrusion.	Dirt entered the ball joint boot and caused premature wear life.	Personnel were reminded that maintenance procedures had to be followed as scheduled.	Closed

Fig. 8.1. Example of typical incidents on a vehicle program with root cause and corrective actions.

Closure of a failure incident has some deep significance. Some unwitting personnel believe that simply changing the failed component with a replacement part fixes the problem. This is not the case, because it does not relate to a long-term corrective action. Only the immediate (remedial) problem was resolved. The design engineer must determine the root cause of the failure. Was it a design problem, a manufacturing or quality anomaly, an operator or maintenance error, an environmental difficulty, or some other causal explanation? The root cause must be ascertained to allow corrective action to be initiated. After implementation, verification must be done to ensure the failure does not recur. The incident can be closed only after some proof of the fix is evidenced.

Data collected during prior testing are useful for the current testing that is occurring. However, gathered information also can be useful for similar or future programs. There should be no question about the data being accurate and detailed enough to enable the analyst to determine the root cause of a problem. The information must be stored and coded in such a manner that it can be retrieved easily and by a specific part number, component, or subsystem. These and other important characteristics must be "called out" and identified. The supplier can utilize a coding system or some other system of retrieval. The important thing is that the routine and specific reports must be obtainable for review and analysis. If the data base is large, it becomes imperative that a computer program be utilized to collect and compile the data.

The code system for classifying failures is basically the responsibility of the manufacturer so that it can establish any effective system for collecting past data. Several factors should be taken into account. The idea is to render the mass of information manageable and understandable to be effective. The coverage used should at least include the following criteria:

- Retrieval by part number
- Retrieval by subsystem code
- Retrieval by failure mode identification
- Status of corrective action (either open or closed)
- Geographic area where the incident occurred
- Time or other measure of the maintenance action
- Time or other measure at the incident
- Environmental conditions before and after the incident

Retrieval by some grouping—such as part number, subsystem code, or failure mode coding—is as important as receiving information of a failure that is under current analysis. Information entered into a databank provides a history of similar parts, related equipment, maintenance times, and other relevant knowledge. Other areas of systematic collection furnish information on location, maintainability problems, and circumstances surrounding the failure. These often provide insight into the current problem and alertness to recurring failures.

When data are collected and organized, the information should be reviewed by a team of design and quality engineers, manufacturing and maintenance personnel, and other concerned individuals. The team often is called a Failure Review Board (FRB). Actions from the FRB or similar empowered team encompass the review of the reported failure or, in some cases, failures of the same type. The failure causes are identified, and the responsibilities for corrective actions are assigned. The FRB should have the responsibility and authority to effectively implement the corrective or preventive action necessary to attain the required reliability. The team should meet regularly, from the concept phase through field testing and the user phase. At times of crisis, the team may meet more frequently.

The overall responsibilities of an FRB generally include the following:

- Evaluating the effectiveness of the system
- Determining the root cause of the problem
- Initiating and then implementing the corrective action
- Correlating laboratory testing results with field results
- Confirming the results of the corrective or preventive action
- Follow-up corrective actions to ensure their effectiveness

Figure 8.2 shows a flowchart of a closed-loop failure reporting system. A customized system also can be fabricated. The flowchart depicts the processing of information from the inception of the failure or incident, to the writing of an incident report, through its analysis and corrective and preventive action. Figure 8.2 can be used for the recommended flow process for a failed component where the failed part must be analyzed for the root cause of the failure. The information derived from the analysis then flows to the organization(s) responsible for the corrective action. The end result should be the corrective and preventive action and implementation of the fix, coupled with the assurance that the incident will not happen again.

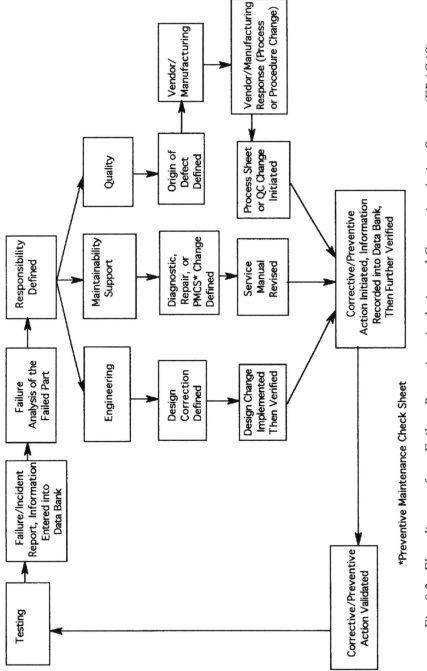

Fig. 8.2. Flow diagram for a Failure Reporting, Analysis, and Corrective Action System (FRACAS).

After a FRACAS has been established, the collective process should be able to retrieve the following information:

- Failure analysis records for specific parts or subsystems
- Individual incident reports
- Group of failures (by code numbers or other identification)
- Maintenance data, including scheduled and unscheduled actions
- Test records
- Types of repairs
- Implemented corrective and preventive actions
- Failure times, miles, cycles, or other measure of performance
- Determine whether incidents are open or closed

Chapter 9 discusses the information that should be found on any failure reporting form. This can include not only a description of the incident or failure information, but also data on the conditions before and during the test, together with other details of what occurred. As in the FRB action, the contents of the form must be reviewed by a team of design engineers and quality, maintenance, and manufacturing personnel to determine the root cause of the failure and then to initiate the corrective and preventive action. If the failure is serious, the user also can have a representative on the team; however, for the general run of failures, the team is composed primarily of the manufacturer's personnel. Often, a failure mode can show that the design was the cause of the failure. At other times, controversy may develop concerning whether the failure is due to a poor design or a manufacturing error. The team approach should resolve the problem, with subsequent implementation of what must be accomplished to correct the anomaly.

Chapter 9 discusses failure analysis and how it relates closely to the FRACAS. Parts and components that fail during R&M testing should be returned to engineering, in order to determine the root cause of the failure. It is not enough to receive a failure report; the failed part also must be returned. The report and the analysis of the failure should provide the design engineer with enough clues to enable him or her to determine what should be accomplished by the corrective action.

Chapter 9 supplies additional details on failure analysis and how it should be achieved. The importance to a FRACAS is that the part must be returned without further damage. Extra care should be exercised during disassembly to retain the state in which the failure of the part occurred.

References

Feigenbaum, Armand V., *Total Quality Control*, 4th ed., McGraw-Hill, New York, 1991.

Lewis, E.E., *Introduction to Reliability Engineering*, 2nd ed., John Wiley and Sons, New York, 1994.

O'Connor, Patrick D.T., *Practical Reliability Engineering*, 2nd ed., John Wiley and Sons, New York, 1985.

Chapter 9

Failure Analysis

Failure Analysis (FA) is more or less a part or extension of the Failure Reporting, Analysis, and Corrective Action System (FRACAS) that was discussed in Chapter 8. Failure analysis probably has been with us since primeval times. Ancient man probably made items that broke. In some crude way, he probably analyzed what went wrong and then either tried to fix the item or make a new one. In some manner, he determined what was wrong, whether it be a club or a spear that had broken. Then, he made a stronger one to ensure that the failure would not occur again. He also had to consider the weight of the club because if it were too heavy, he could not carry it. Conversely, the club would have to be strong enough for him to use as a weapon. Even cavemen had to think about tradeoffs.

Modern man does a similar analysis, but more on a systematic basis. We have failures, in the same way that ancient man had failures. In most cases, we try to determine why and how parts fail. The sophistication can go even further than simply analyzing failed parts. Analysis can be done for potential failures for components that have not yet been made. Chapters 5 and 10, respectively, discuss FMEAs and R&M predictions that fall into the realm of trying to identify and prevent part failures before they occur at the end of their design lives.

This chapter discusses the analysis of failed parts, whether the failures are the aftermath of prototype testing, qualification or acceptance tests, or failures occurring in the hands of the customer. In such cases, the intent is to determine the root cause of the failure and then initiate corrective and preventive action to ensure that the failure does not occur again.

Before discussing the techniques of failure analysis, it is important to understand the definition of a failure. It is surprising at times when one engineer believes a malfunction is not serious enough to be considered a "failure," while another engineer believes the same malfunction definitely is a failure. Usually, functional specification requirements are defined by the customer, and the manufacturer is expected to satisfy those requirements. More simply stated, a failure is an event that makes the equipment deviate from specified limits of useful performance. Taking the definition further, failure is the event that terminates the ability of machinery to perform its intended function. The customer should set the limits of performance and specifications clearly and definably. Often, some marginality may exist in the results of a requirement; however, in those cases, the customer and manufacturer must determine what should be done. At minimum, both parties should know the boundaries of performance.

In general, there are eight types of failure causes:

1. Degradation
2. Gradually developing factor(s)
3. Imperfection
4. Inherent weakness
5. Misuse
6. One-time unrepeatable anomaly
7. Shortcoming
8. Sudden factors(s) that could not be foreseen

An explanation of each of these eight failure causes is probably self-evident, but some interpretation may be helpful.

Degradation is the gradual deterioration or impairment of functionalism. It also can be the process of transition from high to low performance. Sometimes, degradation occurs so slowly that it is hardly noticeable.

Gradually developing factors include longer response times, reduction in protective devices, performance degeneration (similar to degradation), and a loss of design features. Often, such factors are unnoticeable until the performance is not being met.

Imperfection is a withdrawing of a quality characteristic from its intended level, although it is not associated as a specification requirement or its usability. Frequently, this can be a serious annoyance to a customer, even if it is not a contract requirement.

Inherent weakness is a design characteristic that has not yet arrived.

The designer is responsible for assuring that the final product has design maturity. *Misuse* can be an intended or inadvertent maltreatment of the machinery. Sometimes, misuse can extend to the level of abuse. It is difficult to find the root cause of a rare occurrence of a *one-time unrepeatable anomaly*. Nonetheless, such anomalies do occur infrequently. Eventually, the one-time anomaly becomes multiple-time anomalies, and analysis becomes easier.

A *shortcoming* is a design deficiency that has not been uncovered until failure occurs. A shortcoming can be a flaw that is real but tends to be marginal or troublesome while remaining within specification.

Sudden factors are similar, because they too are not anticipated or envisioned.

Most of these failure causes will be discovered through a complete R&M test program. However, it would be much more appropriate if they were found in the concept and early development phases. Chapters 5 and 10 discuss in detail the importance of seeking all potential failure modes before they occur. All design engineers should consider these methods seriously.

Extending causes to the types of failures, where a failure may be any one of the following:

- Catastrophic
- Complete
- Gradual
- Intermittent
- One-shot
- Partial
- Sudden

A *catastrophic failure* is one that is a great and sudden calamity or a disaster. It can be a violent change in which extreme damage is done to the machinery. A catastrophic failure can be one that causes secondary failures within the equipment.

A *complete failure* is one that is one step lower than a catastrophic failure. In a complete failure, damage is done only to the component that failed.

A *gradual failure* is one that advances or progresses in degrees. It usually is noticed first by reduction in performance or output values and then gradually becomes worse.

An *intermittent failure* is one that starts and stops at regular or irregular intervals.

In contrast, the *one-shot* failure occurs only once, without any warning or indication that it is about to happen. When the one-shot failure does takes place, it is final. That is, the failure does not repeat itself.

The *partial* failure does not destroy the total function of the failed part, but it does cause enough damage to make the machinery unable to function as required.

A *sudden* failure occurs without warning and usually is unforeseen. It also is characterized by being abrupt, unexpected, and swift.

Several techniques are available for the performing failure analysis. One method is in the laboratory, where an examination is conducted to determine how a particular part failed. This can be done in the form of analyzing the fatigue marks on a component, or checking the hardness specification, case depth, metallurgical contents, or dimensional characteristics of a part. Use of equipment such as special cameras, x-ray and ultrasound machines, spectrographs, chemical etchings, scanning electronic and optical microscopes, hardness testers, and other more sophisticated instruments can prove useful in the analysis. If the characteristics are within specified tolerances, there may be a necessity to revise the design. The idea is to find the definition of the problem, the mechanics of why and how the problem occurred, or the root cause of the failure.

The *R&M Guideline* discusses Root Cause Analysis, which should target the identification of the basic mechanism by which the problem occurs and should provide a recommendation for corrective action. The interpretation of root cause is acutely important because some technicians believe the changing or repair of the failed part solves the problem. The underlying reason for the failure must be found, not simply the repair of the part. The guideline accepts a "No Corrective Action" decision, but it must be properly justified. To make an honest evaluation of the failure, all reasonable means must be used to determine the root cause. Depending on circumstances, a decision about what should be done may be a tradeoff to which the manufacturer and the user both agree.

The term "random failure" often is used when the failure cause is extremely difficult to determine. Random failures generally fall into one of three categories:

1. A chance failure of a rare statistical occurrence (but if the same failure recurs, the premise becomes weak)

2. A chance failure of exponential equipment such as electronic components that have the same probability of failing for their population (i.e., failures could occur at any time within their distribution pattern)

3. The occasional rare failures that occur between infant mortality and wear-out life. (See Chapter 2 for a discussion of the bathtub curve.) In any event, the failure analyst should utilize his or her expertise to the utmost to determine why a failure occurred. The information is crucial to the designer if he or she is to produce reliable equipment.

The analysis of the failed part usually results in determining the root cause. The following reasons generally relate to the root cause of failure:

1. Accident

2. Design
   - Unprotected areas (e.g., from heat, cold, scuffing)
   - Excessive wear
   - Incorrect inputs
   - Leakage (then determine why the leakage occurred)
   - Mechanical fatigue
   - Metallurgical deficiency (e.g., case hardness and depth, tensile strength)
   - Electrical problems (e.g., shorts, weld or solder joint weakness)
   - Miscalculations in ergonomics
   - Inadequate system integration
   - Material shortcomings

- Inadequate stress/strength determination
- Incompatibility of materials, fluids, metals, etc.
- Software errors

3. Environmental Conditions

   - Chemical imbalance (e.g., attacked by outside substances)
   - Unanticipated shock or vibration
   - Thermal (i.e., an environment that is too hot or too cold)
   - Unexpected conditions

4. Maintenance

   - Misunderstanding of the manual
   - Lack of clarity in repair instructions
   - Use of incorrect tools
   - Maintenance improperly performed per the instructions
   - Lack of maintenance

5. Operator Errors

   - Erroneous usage (misuse)
   - Rough handling
   - Abuse
   - Inadvertent overloading

6. Quality or Manufacturing Errors

   - Dimensional errors usually due to manufacturing flaws
   - Incorrect assembly
   - Use of incorrect parts

Many of the root causes listed here can originate from several sources. For example, metallurgical problems can be either a design anomaly due to the designer not specifying an adequate case hardness and depth, or they can be a heat treat flaw because the part was not heat treated to the correct specification. By the same token, erroneous usage can be misjudgment by an operator, or it can be a repairman using incorrect tools or not following maintenance instructions. Leakage is another example of a design or quality anomaly. A designer may not specify the correct seal for its intended application, or the seal may be the proper design but the production personnel may install it incorrectly. In any event, designers should think about the root causes of potential failures when they design their specific assemblies or components, and then conceive them so that the parts will be adequate for the operations intended. Likewise, production personnel should utilize the necessary care when assembling the parts for which they are responsible. Manufacturing engineers should supply the proper equipment so that the production workers can fulfill their responsibilities.

Chapter 15 discusses some problem-solving tools that also can be used to find root causes. These simple tools of quality sometimes can be extremely helpful when trying to analyze problem areas.

For the QS-9000TE R&M requirements, the failure or nonconformance problem cannot be closed until all required corrective actions have been developed and implemented. All investigations, analyses, analytical results, and corrective actions must be documented and entered into the R&M data collection system. In Chapter 8, Fig. 8.1 showed an example of a series of test results with the root causes and corrective actions that occurred. Notice that the incidents were not closed until a consensus was reached at a design review meeting, in which the particular corrective action was implemented and then was verified over a period of time, miles, or cycles. There must be some indications that the corrective actions will prevent further problems.

Usually during the initial phases of testing, the failure rate of equipment is high, due mostly to infant mortality-type failures of weak parts. The cause of most of these early failures can readily be found and corrected. Most of the infant mortality-type failures are production-type anomalies or immature designs that still have some flaws. To have a comprehensive R&M program, all failed parts and failure reports must be documented and forwarded to the personnel responsible for the design. This can be done in any form the manufacturer deems effective and efficient, but it is crucial that all nonconformances be reported and that all failed parts be returned for analysis.

Upon receipt of the failure information and/or parts, the designer or failure analyst analyzes the failed parts and documentation to determine the root cause, or the reason for the failure. To assist the designer, the following data at minimum should be reported for each failure or downtime incident:

- Facility, plant, or test location
- Fault code
- Initial observation (i.e., description of the root cause if known)
- Time and date of the event
- Time the event cleared
- Environmental conditions
- Repair time
- Duration of downtime
- Total operating time of the product (i.e., overall life of the machine)
- Total run-time of the failed component
- Operation or test number
- Tagging number or coding identification relating to the failure
- Part number of the failed component
- Maintenance times
- Repair or test personnel comments

By coding and retrieving some of these factors, design engineers often can determine trends in failure modes or recognize the number of failures of specific or like parts. The repetition of fault codes, test locations, times of events, environmental conditions, personnel comments, and sometimes even the operator of the equipment frequently can lead to the reasons why a particular machinery fails. The ultimate cause of the failure usually is ascertained by the failure analysis on the parts; however, in many cases, other information can be of assistance.

The *R&M Guideline* provides a "Universal Tag," as shown on Fig. 9.1. This tag also serves as a document to accompany the failed part. It is an ideal identifier to use in reporting and tracking part failures. The depicted tag has many characteristics that are useful not only to the design engineer, but also to manufacturing and quality engineers. It can be used in any convenient format that is agreeable to the customer, supplier, and test facility. Whatever the tag is, it should accompany or, better yet, be attached to the part that failed. At minimum, it should contain the following information:

- Part name, part number, and next higher assembly

- Manufacturer (if known)

- Date of failure

- Test conditions during the time of the failure

- Number of hours, cycles, or miles on the part

- Action taken at the test site

- Opinion of test site personnel about why the part failed

- Any other information that the test site personnel believe would be helpful in the analysis

Fig. 9.1. Universal tag (From R&M *Guideline)*

The reader should note that the information specified on the universal tag is similar to what was requested for each failure or downtime incident. More detail is required for the report than for the tag because of the supporting knowledge it may contain for the analyst.

The important factor is that the information must be readily available to be input into the database, and then be retrievable in an orderly and supporting manner. The test site should be extremely careful in removing the failed part to prevent damage or destruction of any evidence about why the part malfunctioned.

Figure 9.2 shows the flow from the time the failure occurs and the writing of the Universal Tag to the implementation of the corrective and preventive action. The numbers in Fig. 9.2 are keyed to the following processes:

1. The Universal Tag is completed and attached to the part.

2. The information on the tag is entered into a database.

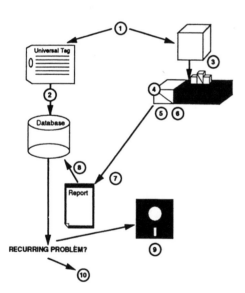

Process Steps:

1. A universal tag is attached to the failed component with a copy to be entered in the plant database.
2. Data is entered in the plant database.
3. A copy of the universal tag is sent to the manufacturing machinery and equipment supplier.
4. If appropriate, the component is sent to the component manufacturer.
5. An evaluation of the component is made to determine root cause of failure.
6. An identification of required corrective action is made.
7. A report is generated by the manufacturer and sent to the user and supplier.
8. The report is entered in the user and/or supplier databases.
9. A report containing all current information and reconciled activity is sent from the user database to the user's plant floor.
10. An exception report indicating recurring failures is generated from the user database. The user and/or supplier team takes action.

Fig. 9.2. Tracking and feedback diagram. (From R&M Guideline*)*

3. A copy of the tag is sent to the machinery and equipment supplier.

4. The component is forwarded to the manufacturer of the part or to a failure laboratory.

5. An analysis is made to determine the root cause of the failure.

6. An identification of the required corrective or preventive action is made.

7. A failure report is generated by the manufacturer or the failure laboratory and is sent to the user and the supplier.

8. Information from the report is entered into the database.

9. This information is sent to the user, assuring him or her that what was done will prevent the failure from recurring.

10. If applicable, an exception report is generated from the database, indicating recurring failures from previous tests. This information is used to analyze the history of the part or similar parts.

The importance of having a failure analysis facility or the availability of a reputable laboratory cannot be overemphasized. The manufacturer must know why a part failed. A competent designer can do only so much; however, a laboratory facility should provide more accurate results. Some of the equipment that should be in a failure analysis facility includes the following:

- Chemical analysis equipment
- Disassembly tools
- Electrical measurement equipment
- Electrical stress equipment
- Electron and optical microscopes
- Fluorescent dye penetrant
- Hardness testers
- Heating oven

- Leak detectors
- Mass spectrometer
- Mechanical shock equipment
- Nondestructive test equipment
- Photographic equipment
- Polarized light
- Radiation type tracer
- Radiographic equipment
- Refrigeration equipment
- Sample fabricator
- Tensile tester
- Ultrasonic equipment
- X-ray radiographic equipment

Such specialized equipment should be found in a laboratory, or the equipment should be available to the analyst who knows and understands how the apparatus is to be used. Some laboratories may require more appropriate or sophisticated equipment; others may require less. Much depends on the type of machinery that is to be analyzed. It is also helpful to know in what state or condition the test machinery was at the time of the failure. The analyst should understand and be able to interpret the meaning of failure analysis terms, so that he or she can explain them to the design engineer if necessary. After the laboratory results determine the cause of the failure, a failure report should be written.

Sometimes, a laboratory analysis may not even be necessary. A cursory review of dimensions, hardness test, or type of fracture tells the reason for the failure. In some organizations, the reliability engineer can perform this service. If the failure analysis appears as if it will be complex, the manufacturer can send the part and documentation to a qualified laboratory.

This chapter discussed the importance of receiving failure information and parts, and determining why a component failed. This is accomplished largely through the formal process of failure analysis. The significance of this process is crucial to properly designing a product because all modes of failures must be detected before the machinery is released for production. This chapter emphasized the importance of finding the root cause and working with the customer. If all is accomplished properly, the end result should be machinery that meets the R&M requirements of the customer.

References

Juran, Joseph M., *Juran's Quality Handbook*, 5th ed., McGraw-Hill, New York, 1999.

O'Connor, Patrick D.T., *Practical Reliability Engineering*, 2nd ed., John Wiley and Sons, New York, 1985.

Reliability and Maintainability Guideline for Manufacturing Machinery and Equipment, Society of Automotive Engineers, Warrendale, PA, and National Center for Manufacturing Sciences, Ann Arbor, MI, 1999.

Chapter 10

Reliability and Maintainability Prediction

Reliability prediction is a technique used to estimate the expected reliability parameters before a part is fabricated. Predictions are based on estimations and on the analysis of the design, or on the performance history of the part. Most predictions are presented in MTBF format or as the probability of survival. The QS-9000TE requires that machine suppliers engage in a comprehensive program to determine the R&M characteristics and/or utilize predictive R&M techniques. It is generally a sound policy for manufacturers to perform preliminary reliability predictions early in the design stage to ensure that bid responses can be supported. Verification of reliability predictions is accomplished when testing is initiated on the product. Feedback and analysis of information demonstrates how well the reliability predictions were made in the first place. As data are collected and analyzed, adjustments are made to the predicted MTBFs and maintainability parameters in order to convey more realistic values.

Maintainability predictions are similar to reliability predictions because they too are used to estimate the expected maintainability parameters before a part is fabricated. Depending on the customer interest and requirements, these can include characteristics such as Maintenance Ratio (MR), Mean Time To Repair (MTTR), Mean Miles Between Preventive Maintenance (MMBPM), Mean Time Between Unscheduled Maintenance (MTBUM), and Mean Time to Perform Preventive Maintenance (MTPPM). Time can be used in the requirements, but miles, cycles, or any other appropriate measure also can be used. The customer defines what he or she requires. The user may stipulate only the reliability parameters and not the maintainability characteristics. If the customer does specify the maintainability values, Chapter 3 describes these and other parameters.

Predictions should be accomplished in a reasonable, scientific manner. Factors such as performance functions, operating environments, use of standard or newly designed parts, relationships among the components, complexity of the assembly, calculations from subcontractors, the use of similar or like parts, and the expected customer utilization all should be evaluated. Reliability prediction often can appear to be a "black art." However, if accomplished in a methodical manner and with suitable analytical thinking, realistic MTBFs can be determined. Predictions generally are the only method, short of actual testing of models, that can determine the reliability of the design.

If an assembly reliability prediction appears to be difficult to determine or too complex to rationalize the relationship, a Reliability Block Diagram (RBD) helps in the analysis. This technique utilizes the estimating of component reliabilities, and then consolidates them into a reliability block diagram and an eventual machine reliability. Care must be exercised in the incorporation of the components because the relationship of mutuality and independence of events must be considered. Figure 10.1 shows a simple RBD for an automotive vehicle.

A cursory review of the RBD or prediction chart shows the system (the total vehicle), subsystems (e.g., power train, electrical, cooling), first-level component reliabilities, Mean Miles Between Failures (MMBFs), and MRs. If desired or required by the customer, other maintainability parameters can be shown on the same chart. If the components all are independent of each other and each of them can fail independently, the reliability of the assembly is the product of all the individual reliabilities (or the components are in a series system). For all intents and purposes, the inverse of the MMBFs or failure ratios is additive if the designer can state that the components follow an exponential distribution. For most components, the assumption is fairly safe, at least for estimating purposes. If the components are not independent of each other, the calculations become more complex.

For the maintainability characteristics, the MRs for the component are directly additive to obtain the system MR. If the Mean Time To Repair (MTTR) is required, the repair times for the subsystems are added for the system being analyzed. Then the sum is divided by the addition of the number of subsystem repairs in the same system.

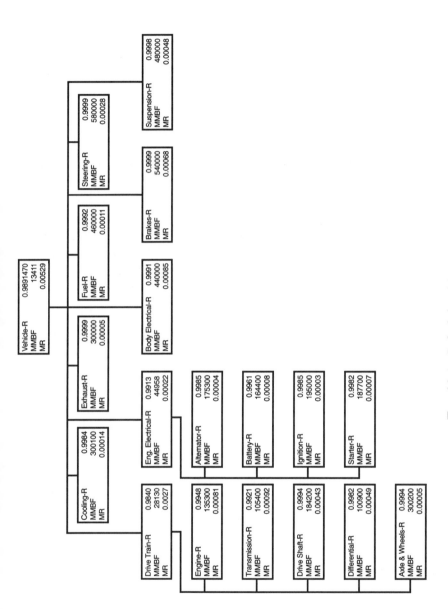

Fig. 10.1. Reliability block diagram.

The following steps provide instructions on how an RBD can be utilized in making reliability predictions.

1. The components, subsystems, and system are defined by their functional relationship. This can be taken down to the lowest-level nut and bolt if desired, but that usually is not necessary unless the customer requires that level of sophistication.

2. The functional relationship is laid out in an RBD. Pay special attention if the relationships are in series, redundant, or in some special complex alliance. The RBD should show the exact functions for the successful operation of the system. It becomes the foundation for developing the probability model for calculating the reliability of the overall system.

3. If the RBD is made of components that are all independent of each other, the individual reliabilities can be multiplied together to determine the system reliability. If the customer requires a specific MTBF, the reciprocals of each component can be added to calculate the system MTBF. By the same logic, failure rate can be added directly.

If component reliabilities are not known, the Society of Automotive Engineers (SAE) has developed part reliabilities for mechanical parts, and MIL-HDBK-217 (Department of Defense) can provide failure rates for electronic components. Values derived from these documents and other sources must be reviewed carefully, as they may not particularly fit the exact functional relationship for what the designer has in mind. Past history sometimes provides a better value. In any event, the design must exercise some positive analytical discretion.

Other sources of reliability information are available:

-   Previous field performance tests that can relate to the designer's product. Always look for like or similar components, and upgrade as necessary.

-   Special life tests on the product or components in question.

-   Other part qualification and inspection test results.

- Government and industry documents—such as PB-181-080 (DOC), MIL-HDBK-217 (DOD), MIL-STD-756 (DOD), 319 (IEC), and Spec. Bulletin 128 (USAF)—are a few publications that can help determine part failure rates and reliabilities.

- Databanks such as the Government Industry Data Exchange Program (GIDEP) and the Mechanical Reliability or Failure Rate Values from the Rome Air Development Center (RADC) can provide in a similar manner useful information to help in the generation of an RBD with predictions.

- If all avenues of predictions have been researched and nothing can be found, "engineering judgment" is a valid method for estimating failure rates. Unfortunately, this depends heavily on the engineer's experience and is considered the least reliable of the prediction methods unless the engineer's background has a tremendous knowledge of the area under analysis.

- Failure Modes and Effects Analysis (FMEA)

- Fault Tree Analysis (FTA)

- Finite Element Analysis (FEA) or similar stress-related analytical techniques often are helpful when reliability information is not available. FEA is especially useful in finding "stress riser" locations when the design engineer can physically revise dimensions or materials to reduce the stress impact.

Reliability allocation (or apportionment because it is another term for allocating) is related closely to reliability prediction. The difference is that prediction is forecasting where the design team calculates what it believes the reliabilities will be, and allocation is what the team believes the reliabilities should be or the goals of the component reliabilities as compared to the whole. Tradeoffs and adjustments often enter the picture to balance factors such as redundancy, costs, safety, and other attributes with reliability. With the TE customer requiring a certain overall reliability or in some cases component reliability, it often becomes a challenge to the designer to match individual reliabilities to achieve system reliability.

Prediction analysis also can be extended to other factors beside R&M parameters. Life Cycle Costs (LCC) can be determined for different cost elements and for total cost of a design. These can be extended for various time periods. Personnel other than cost analysts often are shocked when they discover the overall costing for a piece of machinery. Initial cost of equipment can multiply tenfold when considering the overall cost for a period of time. Chapter 12 discusses the details of LCC.

Monte Carlo modeling is a technique somewhat related to reliability predictions. It is a simulation method for studying system reliabilities by inputting part statistical distribution patterns of their failure models into a computer program. The inputs then are used to calculate system reliability. The method is extremely complex and tedious for manual calculations and thus requires the use of a computer. In effect, the program takes the distribution patterns of the parts that make up the system, randomly selects values from their distributions, and then locates the values into the system distribution after necessary calculations have been made. The system distribution reflects the distribution of reliability estimates. To obtain the system distribution, the program often requires thousands of trials to present meaningful results. Cohen and Nelson provide some additional directions on the use of the Monte Carlo modeling simulation theory.

Maintainability characteristics were discussed previously in this chapter. What is predicted by the manufacturer depends on the parameters required by the user. When actual performance and failure data are obtained from the R&M operations and testing of similar systems over an extended period of time, predictions can prove extremely accurate. This chapter discussed RBD, maintenance characteristics, and the calculation of R&M predictions. Design engineers generally have a difficult time with predictions during the concept phase. However, as the design matures, the quality and quantity of information increase. This scenario, plus the analytical abilities of the team that determines the predictions, thus assists in developing more accurate predictions.

Reliability predictions should be a continuous process. As the development of the product goes through its various stages of reinforcement, data from testing and customer usage should be utilized to update the predictions and apportionments. Although allocations have a tendency to be stationary, a multitude of negative feedback from the customer usually indicates that the product was not adequately designed or that the manufacturing methods require considerable improvement.

Early development and testing must catch all flaws in the system before the product is sold to the customer. Later, validation of the production methods also is necessary, and improvement should be accomplished on a continuous basis where possible. All manufacturers should recognize that costs escalate almost exponentially when the user finds problems. Likewise, there is the forewarning of potential lost sales if failures become prevalent in the field.

References

Cohen, Daniel, I.A., *Introduction to Computer Theory*, John Wiley and Sons, 3rd ed., New York, 1997.

Juran, Joseph M., *Juran's Quality Handbook*, 5th ed., McGraw-Hill, New York, 1999.

Nelson, R.S., *Introduction to Automata*, John Wiley and Sons, New York, 1967.

O'Connor, Patrick D.T., *Practical Reliability Engineering*, 2nd ed., John Wiley and Sons, New York, 1985.

Chapter 11

Machine Qualification and Operation and Reliability Growth Monitoring

The *R&M Guideline* discusses machinery use, machine environment, and performance data feedback in detail. The discussions stress that all machinery suppliers and users should work together and be aware of how the machinery is intended to be used. Factors such as time required to do scheduled maintenance should be considered, along with the accessibility and ease of removing and repairing parts, inspection and diagnostics, and serviceability. Increasing or decreasing speeds and feeds, and the complexity and length of in-line operations and duty cycles also enter the picture of influencing R&M.

In the concept stage, the customer or machinery user specifies the R&M requirements. It is the responsibility of the supplier to address how the R&M requirements will be met and to supply the required R&M values. The TE supplier determines how this is accomplished, but the work should be done in cooperation with the customer. During preliminary design, many considerations and proposals must be evaluated. The coordination of activities starts in the concept stage and can continue through the disposal stage.

Machine users and designers must understand the influences of heat and cold, humidity, fungus, shock and vibration, contamination, salt atmosphere, electromagnetic interferences, and other outside and inside factors. Both customers and manufacturers must know what environmental forces could sway the reliability of the hydraulics, mechanics, pneumatics, electrical, and electronics of the components in the machinery.

Starting with the concept and development stages, there must be some feedback of information on the performance of the machinery. Design engineering personnel must work closely with the manufacturing engineering, maintenance, and supplier individuals. Teams must be organized to involve all relevant disciplines and to ensure that data information is considered as part of the design. Chapter 8 discussed the details of data collection and the Failure Reporting, Analysis, and Corrective Action System (FRACAS). The idea behind acceptable machine operations is that all available information generated during development and testing should be utilized to the maximum extent so that the R&M attributes of the machinery achieve their optimum potentials. Designers must evaluate ease of access, whether modular or singular parts are to be used, what kind of replaceable assemblies are to be considered, and whether strategically placed test points or outside diagnostics should be part of the design.

During the build and install stage, the TE manufacturer should continue to work with the customer. Qualification tests usually are performed by the machine builder, but they also can be performed by the customer, with accompanying feedback of information. Acceptance testing usually is accomplished by the supplier, but this also can be performed by the TE manufacturer. Regardless of whatever agreement is contracted between the supplier and the customer, the feedback of failure information is crucial to the design and development efforts.

In the operation and support phase, the TE manufacturer remains in the picture, although he or she has sold the product to the user. It is important for the manufacturer to receive information on machine operations in order to determine reliability growth and maintainability parameters. Failure reporting is important from a warranty perspective, but it also is essential for servicing, for considerations for improvement, for "what went wrong" in the design and concept stages, and for determining why anomalies were not discovered in the reliability testing.

One of the biggest problems in the build and install stage is control of process variables. This is especially true when production first begins. Infant mortality has a way of appearing early in the installation stage. Potential worker errors can play havoc with production quality and schedules and often are overlooked by designers. Believe it or not, management sometimes can disrupt the R&M process. In its zeal to meet the scheduling requirements of the customer, management occasionally will become inappropriately lax on R&M requirements. The space shuttle *Challenger* launch of 1986, which killed seven astronauts, is a classic example of how management can override the opinions of design engineers.

If manufacturing and quality engineers worked closely with designers in the concept and development stages, and management plays a supporting role, infant mortality failures can be kept to a minimum. Manufacturing engineers should be able to provide some information on machine capabilities crucial to the production process. Consideration for the manufacturing process variables should be identified and controlled long before manufacturing starts.

Some factors that management can control include the following:

- Adequate training of operators

- Machines that are capable of meeting tolerances established by the design

- Proper maintenance on production machines to keep them operating smoothly

- Tooling, fixtures, and gaging are adequate and properly maintained

- Correct materials are being used

- Ensuring that operators examine their work appropriately

- Ensuring that inspectors catch nonconforming components and assemblies

The responsibility of the manufacturing engineer is to assure the design team that these factors are considered long before the design is "locked in." Furthermore, it may be necessary to involve the workers to ensure their understanding of the importance of maintaining the R&M parameters of the product.

As a continuation of machine qualification and operation, reliability growth user involvement during the build and install stage also should be apparent. Potential R&M-related manufacturing problem areas should be investigated, analyzed, and tested if necessary to ensure that the R&M established requirements have a chance of being met. As in the previous stages, the customer and the supplier must work together to determine the R&M baseline.

As a continuation of machine qualification and operation, reliability growth monitoring also becomes part of the picture. Reliability growth management includes the careful identification and cataloging of the successes, failures, and operating times as data is collected from developmental units. Feedback information can be plotted as cumulative data trend lines or as in other methods of charting. The purpose is to determine the status and progress of growth and also to use the chart for future considerations. One item of which the analyst should be aware is to have responsible justification whenever projections are made into the future.

Reliability growth is a good term for the steady improvement of reliability by successful learning or correcting of faults or malfunctions in designed machinery. The maturity of the design can be demonstrated with growth curves that show the reliability of the machinery through the various testing stages. Reliability growth management is accomplished by controlling the failure rate change by learning the factors that affect system effectiveness. These include the refinements of the groups such as the designer, customer, maintenance specialist, manufacturer, and salesperson. System effectiveness should include, but not be limited to, attributes such as:

- Accountability
- Accuracy
- Availability
- Combat effectiveness
- Cost
- Delivery time and method
- Dependability
- Design capability
- Durability
- Environment
- Environmental effects
- Human factors
- Human performance
- Internal availability
- Longevity
- Maintainability
- Mission effectiveness
- Nature of the competition
- Nature of the enemy
- Operational readiness
- Operational readiness constraints
- Performance
- Point availability
- Producibility
- Pseudoreliability
- Quality
- Reliability
- Repairability
- Serviceability
- Standardization
- Steady-state availability
- System reliability
- System worth

Extensive testing usually is necessary to demonstrate reliability growth. Test data in the form of environmental, accelerated, abbreviated, or field testing are utilized to show progress on reliability. If economic constraints do not permit decisive reliability results, the designer may have to depend on analytical methods, some of which may be questionable. Field and customer data are the "proof of the pudding," where data feedback becomes essential to the reliability prediction process.

The essentiality of data feedback, hardware integration, failure reporting and analysis, and customer usage all play a part in reliability growth monitoring. All data should be collected and analyzed. The Duane model is one technique for predicting reliability of future performance from test data and early field data. *Juran's Quality Handbook,* chapter on Product Development (4th edition), shows how the model can be used.

Other methods of demonstrating reliability growth are by fitting a curve to the success-failure data while being cautious about the independence of the test trials and by utilizing maximum likelihood estimates. Lloyd & Lipow go into considerable detail in developing these and other methods of demonstrating reliability growth models.

In the final analysis, the machine supplier is expected to demonstrate the reliability growth of its machinery when required by contractor purchase orders. This can be a part of the continuous improvement process, using all available data in the analysis and test programs. Failure causes must be identified, and corrective actions must be implemented.

References

Juran, Joseph M., *Juran's Quality Handbook,* 4th ed., McGraw-Hill, New York, 1988.

Lloyd, David K., and Lipow, Myron, *Reliability: Management, Methods, and Mathematics*, Prentice-Hall, Englewood Cliffs, NJ, 1962.

Chapter 12

Life Cycle Cost

The Life Cycle Cost (LCC) of a product is the sum of the acquisition costs and logistic support or operating cost in the sequence of phases through which the machinery or equipment passes from conception through decommission. The *R&M Guideline* divides the cost into two main categories: the nonrecurring cost and the support cost. The LCC can include the initial cost of development, plus all other costs associated with the product. Before we discuss how the *R&M Guideline* develops the determination of LCC, it is important to understand the concept of Quality Cost as it relates to QS-9000 and the TE Supplement. The *R&M Guideline* does not specify Quality Cost *per se* but implies it in several areas. Quality Costs are an unquestionable part of the LCC and should be considered in the LCC summation. The following sections elaborate on this point and show the importance of Quality Costs and why manufacturers should consider it as a part of LCC.

The first bullet point of Paragraph 4.2.3.1, Advanced Product Quality Planning, of QS-9000 states:

> The suppliers shall utilize an advanced quality planning process, embracing reliability and maintainability through the Life Cycle Process.

This is a very short sentence for a tremendous amount of effort. The LCC requirement is specified again in four other elements of the *TE Supplement*. As in most other requirements in the *TE Supplement*, the supplier permits the manufacturer to utilize the method(s) that the manufacturer believes to be best for accomplishing this.

The *TE Supplement* offers a simple definition of LCC:

> The sum of all cost factors incurred during the expected life of the machinery.

The Life Cycle is further defined as

> The sequence through which machinery and equipment passes
> from concept through decommission.

This means that the manufacturer (and possibly the user) first should think about the machinery in the concept phase, then continually through all phases of development, testing, manufacturing, and all other costs associated with the product, finally to the customer who uses the equipment, and in the end may decommission it.

Manufacturers may believe LCC is a new concept that is being required; however, it has been in existence for at least four decades but under a different name. Dr. Armand Feigenbaum (b. 1920) initiated the notion of quality costs in the early 1950s, defining the importance of this development to suppliers and to customers. Feigenbaum emphasized that the producer had to understand that a satisfactory product and service quality went hand-in-hand with the satisfactory product and service cost. Similar to several other "quality gurus," he belittled the concept that achievement of better quality required higher costs. Manufacturers today are beginning to understand that if they realize where their costs are, they can do something about managing and controlling those costs.

Feigenbaum discusses "cost of control" and "cost of failure of control." The cost of control breaks down into prevention costs and appraisal costs. The cost of failure of control segments into internal failure costs and external failure costs. Logic tells us that if prevention costs are increased to pay for the correct kind of system engineering work, there would be a significant reduction in the number of defects and nonconformances, lower inspection costs and test activity costs, and consequently higher product reliability. The end result is a general reduction in overall costs, an increase in the level of reliability (and decreased necessary maintainability), and a more satisfied customer.

Feigenbaum goes into considerable detail in identifying the cost factors, such as collecting and reporting quality-cost information, analysis, goals, return on investment, supplier costs, and other costings that are related to LCC in one way or another. He defines the quality-cost items as the cost of prevention that includes quality planning, process control, design and development, quality training, product design verification, systems development and management, and administrative costs. Much of these cost items are referred to in the *TE Supplement* but under different terminology.

Cost of appraisal includes test and inspection of purchased material, laboratory acceptance testing and other measurement services, inspection, testing, checking labor setup for test or inspection, test measuring equipment, quality audits, outside endorsements, maintenance and calibration of test and inspection equipment, product engineering reviews, and field testing. These are mentioned only briefly in the *TE Supplement*.

Cost of internal failures includes scrap, rework, material-procurement cost, and factory contract engineering. Cost of external failures includes warranty, product service, product liability, and product recall. All these are paramount cost factors that contribute to the overall cost of the machinery.

Feigenbaum does not use the term LCC; however, he covers the subject in abundant detail. Many of his terms associate directly with those used in the *TE Supplement*. The specification for maintenance, service, repair, and replacement parts should be considered. Statistical sampling and cost data collection control, management, training, electronic data processing, determination of report formatting, review of trends, the implementation of corrective and preventive actions, and the establishment of quality audit activities all enter the domain of LCC.

QS-9000 does not cover much about LCC; however, the implication of quality costing is in most of the elements. Most manufacturers seeking QS-9000TE compliance strive for the QS-9000 certification. With this in mind, it is a good idea to interpret the elements of QS-9000 as they relate to quality costing.

Element 4.1, Management Responsibility, discusses prevention of nonconformances, the correction of quality problems, setting of objectives and organizational structure, defining responsibilities and authorities, and other requirements related to quality costs. Preventive measures are involved in most of Element 4.2, Quality System. A few of these include the preparation of procedures to ensure personnel knowledge of working conditions, the establishment of a quality plan, the product approval process, continuous improvement, and facilities and tooling management. All of these conditions and situations presume that nonconformances occur and that organizations should be thinking about methods to prevent discrepancies wherever possible. Element 4.3, Contract Review, and Element 4.5, Document and Data Control, are very similar. If purchasing did not concern itself with a clear definition of what the manufacturer plans to buy, we can imagine how the cost of quality would rise. The receipt of nonconforming material due to poorly descriptive specifications can play havoc in any organization. The major measure of Document and Data Control is to have procedures to control quality costs.

The review of the other elements of QS-9000 are similar. Related to appraisal costs are Element 4.9, Process Control; Element 4.10, Inspection and Testing; Element 4.11, Control of Inspection, Measuring, and Test Equipment; and Element 4.12, Inspection and Test Status. Element 4.9, Process Control, also ties in with preventive costs because if parts are not made to specification in the first place, additional costs definitely enter the picture. Checking quality is a crucial criterion; however, if parts are fabricated consistently within required tolerances, the appraisal cost becomes minimal. The QS-9000 procedures theoretically should lead an organization to that premise.

Element 4.13, Control of Nonconforming Product, and Element 4.14, Corrective and Preventive Action, speak for themselves. Both elements deal with internal and external failures and actions that should be taken to prevent nonconformances in the first place, or what can be accomplished to prevent failures from recurring. Element 4.15, Handling, Storage, Packing, Preservation, and Delivery, covers preventive cost. Anything that can be achieved in these areas will help to lessen quality costs downstream.

Some may debate the category in which Element 4.17, Internal Quality Audits, falls; nonetheless, it is a part of quality costs. In one sense, quality audits relate to appraisal. In another sense, it can be prevention also. The manufacturer can place this element where he or she thinks it should be placed. Element 4.18, Training, and Element 4.20, Statistical Techniques, with regard to quality systems both relate to prevention. If servicing relates to the failure of machinery in the hands of the user, it is an external failure cost. If the contract relates to a maintenance clause(s), the cost relates to prevention. Each manufacturer should know how all these costs relate to the four categories of Quality Cost.

Now that we have shown how Quality Costs are part of the LCC, let us return to the *R&M Guideline* that appropriately identifies the life cycle phases of machinery and equipment. However, the *R&M Guideline* fails to include the details of the Quality Cost associated with the LCC. The following is a breakdown of the LCC through the sequence of phases.

1. System Concept and Definition ────────┐
2. Design and Development ────────────┤──── Nonrecurring Cost
3. Manufacturing, Building, and Installation ──┘

4. Operation/Support ──────────────┐
5. Conversion/Decommission ──────────┤──── Support Cost

These in turn are broken down further into subgroup classifications to ensure that the detail costings are accounted for and how they are involved. The total LCC consists of the following relationships:

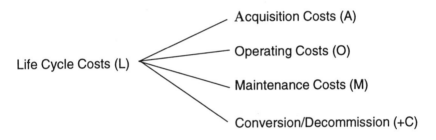

Life Cycle Costs (L)
- Acquisition Costs (A)
- Operating Costs (O)
- Maintenance Costs (M)
- Conversion/Decommission (+C)

where

$$L = A + O + M + C$$

The Acquisition Costs (A) are determined using the following relationships:

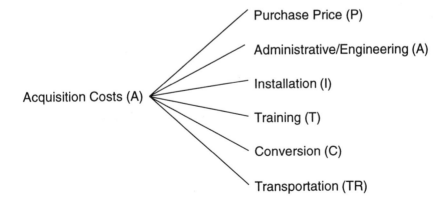

Acquisition Costs (A)
- Purchase Price (P)
- Administrative/Engineering (A)
- Installation (I)
- Training (T)
- Conversion (C)
- Transportation (TR)

where

$$A = P + A + I + T + C + TR$$

The Operating Costs (O) are determined as follows:

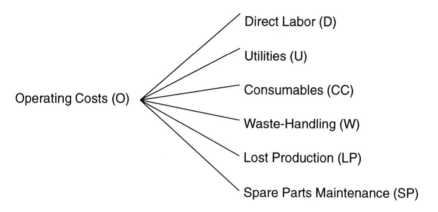

where

$$O = D + U + CC + W + LP + SP$$

Maintenance Costs (M) are determined with the following relationships:

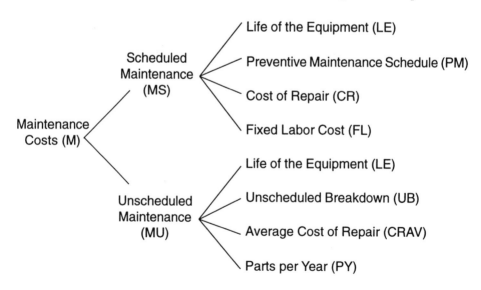

where

$$M = MS + MU$$
$$MS = LE + PM + CR = FL$$
$$MU = LE + UB + CRAV + PY$$

As mentioned in other chapters, management support for the R&M program is essential. If such support does not exist, designers will find it extremely difficult to meet customer requirements. One of the major concerns during a development program is to keep cost low. This is true not only from a profit-making viewpoint, but also to be competitive. Life Cycle Cost is not only of interest to management. It is one of the top priorities of the customer.

What are the responsibilities of management during the five phases of the management process and the five phases of "Manufacturing Machinery and Equipment Life Cycle"? Following are the five phases, accompanied by a brief description of the activities and the management task to support the process.

Phase 1—Concept

The first phase is research and early development or design, usually accompanied by a proposal. The user and TE supplier work together to establish the system requirements. The user team consists of machine operators, engineering staff, and maintenance personnel. The supplier team includes staff from design, engineering, quality, cognizant shop personnel who are aware of machining cost, service, and any subcontractors who will be involved. The combined team usually discusses the machinery mission, the R&M and LCC requirements or goals depending on the contract, safety issues, and quality and environmental concerns. There should be understanding and agreement on the terms, specifications, capacities, the R&M (i.e., firm or target values), and how the concepts can be met. The teams should agree on the following:

- Safety requirements

- How parts are to be controlled

- The basic maintenance approach and philosophy

- The environmental factors involved

- How data will be collected, controlled, and analyzed, which further includes:

   Documentation to support the collection and analysis
   Type of logistic support
   Proposed inspection and testing
   Concerns or constraints

All these concerns may not be agreed to in the concept phase and may be omitted or agreed to in Phase 2. Understandings must be reached between the customer and supplier and then specified in the contract language.

Phase 2—Development and Design

The second phase solidifies the issues of the concept phase and determines most of the LCC. The major design and development characteristic become part of the design. These include R&M, safety, ergonomics, technical and quality requirements, and maintenance parameters. Also considered and finalized are other support activities. These include documents needed; prediction techniques; demonstration, quality, and test requirements; schedule of design reviews; type of data collection and analysis; and selection of subcontractors. By now, purchasing staff should have a good idea about who the suppliers will be. Some of these actions may require negotiation. However, by the time Phase 3 starts, these issues should be resolved.

Phase 3—Build and Installation

Now the manufacturer is beginning production of the product. The qualification test that was fabricated with production tooling should have been completed or be nearing completion. Quality personnel should know what to look for and how to inspect incoming, in-process, and final components and assemblies. There should be assurance that parts are manufactured to design tolerances and are functioning properly. The requirements of QS-9000 elements should be in place and applied as applicable. Both the manufacturer and the user monitor the R&M and ensure that any problems that arise are managed and controlled. Acceptance testing usually is initiated by random selection of the early units that are assembled in the production schedule. Other important activities include the development of maintenance, start of necessary training, identification of assembly and reassembly problems via teardown studies, and debugging of infant mortality-types of failure. Data collection, FRACAS, and R&M demonstration should be well in place, and positive results should be obvious in corrective and preventive action activities.

Phase 4—Operation and Support

The user should have some of the early production units at the start of Phase 4. All setup activities in the earlier phases should be well in place and functioning smoothly. Quality and service personnel should be expected to assist in operation and support where necessary. Although acceptance tests may be in progress, the user can perform his or her own unofficial examinations and report the results to the supplier. Both the user and the manufacturer should be exchanging R&M information and support data continuously.

Phase 5—Conversion and/or Decommission

This is the end of the expected life of the product. Depending on the results of prior phases and what is occurring in Phase 5, the user and the supplier may have to negotiate terms of the contract. Under the worst scenario, rebuilding or decommissioning may occur. If all requirements have been met, that is an excellent condition. However, even if this does not happen, continuous improvement should remain in the activities of both the supplier and the user. Likewise, both also should continue to exchange R&M information.

Summary

A summary of the phases was expanded from the *R&M Guideline*:

Generic R&M Program Matrix

Reliability Requirements	x			
Maintainability Requirements	x			
Failure Definition	x			Concept—Phase 1
Quality Requirements	x			
Environment/Usage	x			
Concept Review	x			
Design Margins		x		
Maintainability Design and Predictions		x		Design and
Reliability Predictions		x		Development—
FMEA/FTA		x		Phase 2
Verification of Quality Requirements		x		
Development and Design Reviews		x		
Parts			x	
Tolerance Studies			x	Build and
Stress Analysis			x	Installation—
Reliability Qualification Testing			x	Phase 3
Reliability Acceptance Testing			x	
Build and Installation Review			x	
Reliability Growth and Maintainability Improvement				x
FRACAS				x O & S
Continuous Improvement				x Phase 4
User Demonstration				x

144

Suppliers and customers should keep in mind that what is described here is what usually occurs in their relationship during the LCC phases of a contract. The activities can be varied as agreed among the negotiating groups, but the recording of R&M performances and the commitment to FRACAS responsibility are essential to any successful program. The commitments and responsibilities must be in place and be pursued energetically.

References

Dhillon, Balvir S., and Reiche, Hans, *Reliability and Maintainability Management*, Van Nostrand Reinhold Co., New York, 1985.

Feigenbaum, Armand V., *Total Quality Control*, 4th ed., McGraw-Hill, New York, 1991.

Jones, James V., *Engineering Design Reliability, Maintainability, and Testability*, Tab Professional and Reference Books, Blue Ridge Summit, PA, 1988.

Juran, Joseph M., *Juran's Quality Handbook*, 5th ed., McGraw-Hill, New York, 1999.

Reliability and Maintainability Guideline for Manufacturing Machinery and Equipment, Society of Automotive Engineers, Warrendale, PA, and National Center for Manufacturing Sciences, Ann Arbor, MI, 1999.

Chapter 13

Diagnostics

Diagnostics is the technique by which equipment malfunctions or processing errors can be detected, discovered, or further isolated. It is the process of studying symptoms, theorizing about causes, testing the theories, and discovering the reasons for nonconformances. The diagnostic examinations can be accomplished manually or automatically to isolate the malfunction. The equipment can have a simple visual "go/no go" display or a complicated software program. In some sense, the definition can be extended to provide for what the necessary equipment and instructions are, and for the repair of the particular malfunction or error.

The technique usually is conducted with the use of special equipment that is either built-in or external to the machine that is under test. The equipment sequentially checks or locates the problem areas either automatically or in manual check-off points. This is called a debugging routine or malfunction routine. Depending on the complexity and function of the equipment, the diagnostics often can indicate possible faults in not only the machinery that is being tested, but also on interfacing components that affect the functioning of the machinery.

The *R&M Guideline* offers one way a diagnostic device can be utilized. The type described is Built-In Test Equipment (BITE) that physically is part of the machinery and indicates the status of the product. In some cases, the diagnostic equipment specifies what parts are to be replaced in the failed unit. The procedure can be part of a diagnostic routine that usually includes a logical sequence of checks or tests to locate the malfunction.

Another type of diagnostic equipment is external to the machine. This type is used similarly to BITE, except that it is utilized after an anomaly has occurred. When a diagnostician wishes to determine the cause of the problem, he or she utilizes an external device designed specifically for the machinery and usually finds that the device is quite helpful.

The diagnostics themselves can be simple visual displays that indicate the status of the machinery by means of an attribute signal showing a "go/no go" condition. This can be in the form of a flashing light, a noise indicator, or any other signal, suggesting a problem or a warning of a potential problem that may be present (e.g., the warning light in your car tells you that the oil pressure in your car is low). Other devices can be a complicated signal that only a well-trained technician can interpret and analyze. In most cases, only the trained technician can recommend the corrective action that should be taken. An x-ray technician is a good example of what a trained individual should be. This is true for both medical and metallurgical examiners.

One of the features of diagnostic systems is that some designs have the capability of storing equipment performance data as a permanent record for analysis and feedback of information. It also is helpful if the output from the diagnostic systems is in a format compatible with commercially available database management software.

Thinking about how the diagnostics should operate starts in the concept phase of a product. This is as true in diagnostics as it is in the other R&M disciplines. Component assemblies and subsystems should have the necessary hardware and software in place and should be capable of supporting the reliability growth management process. By the build phase, the machinery should have proven working units recording and analyzing as required. The units should have the capability to indicate the specific components that must be repaired or replaced. Output from a diagnostic system should be in a format that is compatible with other commercial software.

Training should include the use and understanding of BITE, which is a unit that is a part of the system and an identifiable internal unit of the system. The BITE is used for the specific purpose of testing the machinery. This can be done automatically or on call. Built-In Test (BIT) is another reference to BITE, which usually is composed of self-test hardware and software. The icon lights that flash on the dashboard of your automobile are good examples of BITE.

The responsibility for diagnosis can vary, depending on the size and organizational setup of a company and the complexity of the equipment it produces. Specifically trained diagnosticians can perform the analysis and repairs, although diagnostic teams are capable of doing the same type of function. Sometimes, this is essential, depending on the complexity of the machinery. In any event, someone or a team should be designated to find the cause of the malfunction or failure. A qualified technician must understand and be knowledgeable about the equipment he or she is using and know how to decipher its readings. The designer must relate to what has been discovered, then initiate the corrective action to prevent the malfunction from recurring. In some cases, the equipment specifies what parts must be replaced. It is essential to isolate exactly why a particular machine failed and then correct the specific problem to ensure that the problem does not happen again.

From a broad perspective, the malfunction is diagnosed and solved in the following steps:

- A description of the problem is documented.

- Note when the problem first occurred and what occurred in subsequent stages.

- Determine what could have been done to detect the problem earlier or to prevent the problem from occurring at all.

- The failed machine is diagnosed by a diagnostician or team to determine the cause of the malfunction (if this has not already been accomplished).

- The designer analyzes what the corrective action is and takes the necessary steps to prevent recurrence of the problem.

Training is as essential for diagnostics as it is for many of the other necessary skills of R&M. If a diagnostician does not know how to use the equipment for which he is responsible, the equipment basically is useless. The level of training depends on the complexity of the diagnostic device and what is involved in its relationship with the machinery it is supposed to analyze.

Instructions accompanying machinery that has BITE should be easily understood and well tested before release to the customer. Each signal, analog, or digital readout should have a meaning, and the operator should know what he or she must do to ensure that the machinery remains operational, or if it is in a potential anomaly mode.

149

Diagnostic equipment external to the machine is called a Diagnostic Service Kit (or Equipment) and also can be part of an R&M concern. Similar to BITE, the equipment should reliably detect and isolate specific malfunctions or errors. Similar to BITE, diagnostic kits usually are used by a diagnostician or maintenance specialist to determine the problem and what must be accomplished to fix the difficulty.

In summary, the design engineer should consider making diagnostic equipment part of the machinery he or she is developing. If not internal to the machinery, then there must be some equipment for external support if the BIT is not practical. We have seen that the equipment can be simple or sophisticated. In any event, we must recognize that personnel using the diagnostics devices understands the workings and interpretations of those diagnostic devices. Data from these devices should be permanent and should be available for supplier feedback if required. Manufacturers should realize that training will be necessary for personnel who will operate the diagnostic equipment. Overall, the designer has responsibility for ensuring that the equipment becomes a part of the R&M process and that operators know and understand the devices.

References

Feigenbaum, Armand V., *Total Quality Control*, 4th ed., McGraw-Hill, New York, 1991.

Juran, Joseph M., *Juran's Quality Handbook*, 5th ed., McGraw-Hill, New York, 1999.

Reliability and Maintainability Guideline for Manufacturing Machinery and Equipment, Society of Automotive Engineers, Warrendale, PA, and National Center for Manufacturing Sciences, Ann Arbor, MI, 1999.

Chapter 14

Reliability Improvement Methods

Machine manufacturers should always be conscious of continuous improvement. This is true not only in the analysis and correction of all failures, but also in the proactive mode of thinking that things can always be improved. Designers are not the only individuals who should contemplate reliability enhancement. Everyone in an organization should do this. The following are several areas in which improvements can be initiated to enhance the overall R&M of machinery. Manufacturers also should be able to use at least some of these methods to assist them in the continuous improvement process.

Benchmarking

This tool is not often considered as being a reliability improvement technique. However, searching for the best-in-class and innovatively improving the product certainly places the approach into an enhancement method. Chapter 2 provides additional insight into how the routine is utilized.

Derating

This is the upgrading of component reliability by intentionally reducing the stress/ strength level in the application of a component or increasing the strength characteristics of the part. Usually, this limits the stress level in the part and consequently improves its reliability. A few examples include:

- Use equipment continuously rather than in cycles.

- Use higher-voltage capacitors.

- Use higher-wattage resistors.

- Reinforce weak areas.

- Reduce operating temperatures (see "Operating Environment Control" later in this chapter).

- Reduce capacitance requirements.

- Use strong springs or more durable components.

- Spread out high-stress areas.

Diagnostics

Diagnostics is a procedure used to detect, discover, or further isolate an equipment malfunction or a processing error. Tooling and equipment suppliers and manufacturers should consider the means by which diagnostic measures can be utilized to detect errors. This can be accomplished by computer programs or by mechanical or electronic means. Diagnostics can be built-in or external to the products that have failed. The purpose of diagnostics is to have a systemic approach to finding the cause of either a problem or a potential problem that is ready to happen. Ideally, the diagnosis identifies a symptom, finds the cause and analyzes it, and provides some form of remedy. Chapter 13 discussed diagnostics in additional detail.

Durability

As discussed in previous chapters, durability is a form of reliability. Durability refers to the life of the machinery that can be repaired, serviced, overhauled, or rebuilt. (For example, a light bulb cannot be repaired after it has burned out.) In some sense, if rebuilding can be accomplished at an economic level, the reliability of the equipment can be extended. The design engineer should keep this approach in mind when he or she is planning the machinery.

Fail Safe

In the fail safe method, the failure of one part does not cause the failure of another part. Some engineers also refer to this characteristic as one that prevents harm to humans or social damage due to a faulty system.

Failure Review Board

A Failure Review Board (FRB) is a team of representative personnel from appropriate organizations, who have responsibilities and authorities to ensure that failure causes are identified and that corrective and preventive actions are initiated and validated. Members generally are from field engineering, manufacturing, quality, design, marketing, and project management. However, they also can include others as needed. Starting in the concept phase, the team meets regularly to discuss, review, and initiate the actions necessary to achieve system effectiveness and the R&M characteristics specified by the customer. Additional details about Failure Review Boards are provided in Chapter 8.

Finite Element Analysis

Finite Element Analysis (FEA) as a technique is not considered a reliability improvement method, but it can contribute significantly to the enhancement of reliability. Finite Element Analysis is a technique of modeling a complex structure into a collection of structural elements that are interconnected at a given number of nodes. The model is subjected to known loads, whereby the displacement of the structure can be determined through a set of mathematical equations that account for the element interactions.

Function Analysis

Review of customer usage of the machine parts is known as function analysis. This is especially important if parts fail or are not really necessary for the overall function of the assembly. If a part is not functionally essential to the operation of a machine, eliminate the part.

Idiot Proofing, or Mistake Proofing (Poka-Yoke)

Mistake proofing as a method of enhancing reliability sounds simple, but how often does a machine design allow mistakes or misapplications to happen? A more harsh term is "idiot proofing," which states basically the same principle but does not leave room for an operator to use the equipment improperly. A governor on a machine is a good example of idiot proofing. The mechanism does not permit the machinery to exceed the intended boundary of operation. Other mistake-proofing components and overall design error controls should be considered when designing equipment.

Limited Life Components

Some parts can operate for designated hours of longevity or some other measure of the life of the part. However, some parts must be replaced periodically. The maintenance manual generally is an ideal place to identify what parts must be replaced because the component in question has a defined service life under certain operating conditions.

Link and Interface Analyses

This is the study of linkages and interfaces between and among components. Links are any connections among components, persons, controls, displays, stations, etc. which often are forgotten when analyzing component R&M. Designers also should think about the complete system when they analyze the connections for adequacy, loading, position, arrangement, and cycling. Interface analysis is similar to link analysis in some respects. It also is the study of relationships among components where it seeks any incompatibilities that may cause a problem, especially those that could be a potential hazard. Clearances and mismatching between inputs and outputs are attributes to look for.

Maintenance

Maintenance is work performed to maintain machinery and equipment in its original operating condition to the greatest extent possible. It includes scheduled and unscheduled maintenance activities, but it does not include minor construction or change work. Designers must include maintenance procedures, schedules, and resources as part of the design. Customers should be expected to follow the supplied directions, including replacement of low-reliability components whenever restoration is specified.

Operating Environment Control

This is the constraining environmental influence on a component. Machine suppliers should consider shields, parts that require protective coatings, or a cooling supply to components that are vulnerable to extreme heating. Other areas that may require protection and consideration include:

- Differential pressures
- Extreme dryness or extreme moisture
- Solar deterioration (i.e., radiation and ultraviolet rays)
- Wind damage
- Smog
- Salt spray
- Vibration and shock
- Corrosion
- Terrain

Redundancy

Redundancy refers to providing more than one means of performing a function. Machine suppliers may be able to use this enhancement technique to increase the reliability of their designs. Sometimes, this can be expensive; however, if any electronic components are involved, the improvement in reliability can outweigh the cost. Four basic steps are necessary when considering a robust design:

1. List the important quality characteristics, sources of noise (e.g., variable environmental conditions, raw materials, and processing variables), and control parameters.

2. Plan some tests to determine the effects of the noise.

3. Run the tests, trying to identify the critical control parameters and optimum settings.

4. Run additional tests to confirm and verify what was discovered.

Research and Development

Research and Development (R&D) attempts to increase reliability with new parts, new materials, better methods, or other means of advancing state-of-the-art technology. As in the other processes of continuous improvement, R&D is another mode of operation TE suppliers can utilize to enhance reliability. In the R&D area, the designer must understand and test new methods that are proposed. Experiments often must be conducted, and meticulous records must be maintained.

Robust Design

This phrase refers to designing components so that their manufacturing is relatively simple and that they can manage unexpected environments. Machine builders should keep this cognition in mind, as well as suppliers who do their own designs. Manufacturing engineers should work closely with designers to develop robust designs.

Sacrificial Parts Usage

Some designs incorporate parts that are expected to fail when overload or abuse conditions occur on the machinery. Devices such as fuses, fusible links, shock absorbers, and shear pins are designed to fail before major damage happens to the components that the devices are expected to protect. In some sense, they also may function as safety devices. However, in the long run, they enhance the overall reliability of the machinery. Also, it is a good idea to point out these devices in the operating and maintenance manuals to ensure that personnel are aware of the presence of such devices.

Safety

This factor is essential for reliability. If machinery cannot be operated safely, reliability has little meaning. Design engineers must work closely with safety engineers to ensure that safety is not compromised for reliability. Chapter 16 provides further details about safety concerns.

Screening Tests

Tests that are used to detect early or infant mortality-type failures are known as screening tests. Tooling and equipment suppliers can use this technique in the form of accelerated testing, bench testing, "burn-in tests," or "overstress" tests. These tests are conducted to identify early failures due to weak parts, workmanship defects, and effects of extreme conditions. Sometimes, stress levels higher than those specified are used to test the parts. When doing accelerated tests, the designer must correlate the accelerated time performance against the normal use of the product. This can be tricky at times.

Simplicity

In this complex world of high-tech machines, electronic gadgetry, and multiple purposes, design engineers should keep the KISS principle in mind: Keep It Simple, Stupid. This principle can be true not only for the design of the machine, but also for the operation of the equipment. Generally, fewer parts in an assembly results in higher reliability.

Sneak-Circuit Analysis

This computer-aided technique identifies latent paths in a circuit which could cause problems. To do this analysis, all components should be functioning properly. Conditions for which the designer should look include:

- Improper connections
- Operation error due to lack of instruction or direction
- Circuit paths that were not contemplated

- Software problems caused by incompatible interferences
- Inappropriate timing sequence

Standardization

Standardization refers to the use of proven, standard items. This can be in the form of off-the-shelf items or parts that have been sustained in former designs. Caution should be exercised to ensure that stress levels are compatible with the part. However, in most cases, the standardization of parts generally is a support method of enhancing reliability. Approved parts lists and critical components lists often are useful when considering this method.

Tradeoffs

The optimum and most beneficial balance between two or more interdependent system characteristics can be described as a tradeoff. Tooling and equipment suppliers should work closely with their customers to determine what characteristics are the most important with regard to overall reliability considerations. Both TE suppliers and customers should evaluate their designs with regard to the ability of those designs to meet end-item performance, as well as costing and potential scheduling problems. The designers must analyze alternatives and design to the optimum condition. Areas that can be considered in a tradeoff include:

- Cost of the design
- Cost of product improvements
- Cost of redundancy
- Cost of service
- Total cost
- Manufacturing lead times
- Maintainability schedules

- Performance
- Reliability
- Sample size
- Service provisions
- Statistical confidence
- Supportability
- Test time

Worst Case Analysis

This analysis involves verifying a detailed electrical circuit in conjunction with a computer program. This is especially important for machine builders who use electrical circuits or computers as part of an overall system. The designer should consider "worst case" parameters when specifying the reliability tests that are to be performed. Associated with worst case analysis is "worst case condition." This is the simultaneous application of the most hostile conditions to a test situation to determine how the product would be affected. It is important for the designer to know and understand how the most stressful conditions would affect the performance of the product.

Summary

A review of the improvement methods shows that many of these methods are related to each another. The overall idea is to continually ask the question, "How can I improve the product?" Use the appropriate method that fits the situation and application.

The methods depicted here offer guidance and ideas on how potential improvements can be initiated. Other chapters of this book discuss the use of FMEA, FTA, FRACAS, reliability predictions, and other methods, all of which contribute to the improvement process. These separate chapters present details on how these other techniques can be utilized.

References

Feigenbaum, Armand V., *Total Quality Control*, 4th ed., McGraw-Hill, New York, 1991.

Jones, James V., *Engineering Design Reliability, Maintainability, and Testability*, Tab Professional and Reference Books, Blue Ridge Summit, PA, 1988.

Juran, Joseph M., *Juran's Quality Handbook*, 5th ed., McGraw-Hill, New York, 1999.

Lewis, E.E., *Introduction to Reliability Engineering*, 2nd ed., John Wiley and Sons, New York, 1994.

Lloyd, David K., and Lipow, Myron, *Reliability: Management, Methods, and Mathematics*, Prentice-Hall, Englewood Cliffs, NJ, 1962.

O'Connor, Patrick D.T., *Practical Reliability Engineering*, 2nd ed., John Wiley and Sons, New York, 1985.

Reliability and Maintainability Guideline for Manufacturing Machinery and Equipment, Society of Automotive Engineers, Warrendale, PA, and National Center for Manufacturing Sciences, Ann Arbor, MI, 1999.

Chapter 15

Problem-Solving Techniques

In the R&M process, data can be analyzed in many ways. The choice of how to analyze problems and incoming information generally is determined by the individual performing the analysis. Because of the importance of problem-solving techniques, this chapter describes some of the most prevalent methods that assist in R&M engineering problem solving and solutions.

Seven quality tools initially are discussed, followed by other useful methods. The seven original tools include:

1. Scatter diagram
2. Histogram
3. Flow chart
4. Cause and effect diagram
5. Pareto chart
6. Control chart
7. Check sheet

These seven tools generally are accepted as the seven tools of quality, although some authors replace them with other tools, including (but not limited to) the following six tools:

1. Brainstorming
2. Stratification
3. Run charts or trend charts
4. Nominal grouping technique
5. Force field analysis
6. Process capability

The latter part of this chapter discusses the seven management tools. The seven quality tools are more appropriate for data collection and analysis. The seven management tools are intended for planning, identifying, and coordinating, but they also can be used for data collection and analysis. All of the tools are appropriate for problem solving or for managing specific projects.

Explanations of how the tools can be used and what results can be expected are provided in the following sections. The tools are described briefly, and a few examples are provided.

Scatter Diagram

One of the simplest of the quality tools is the scatter diagram, which pictorially presents on a graphic form the relationship between two variables. Figure 15.1 shows the relationship between torque readings before and after vibration tests. The chart depicts a scatter of points about a regression line that is in the positive direction, i.e., as the before-vibration torque increases, the after-vibration torque also increases but not on a 45° relationship. A negative trend would indicate an increase in one variable and a decrease in the other. An example of a negative trend is that as the quality of a product increases, the warranty cost on the same product decreases.

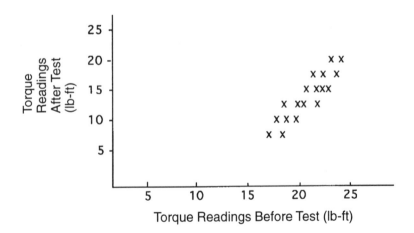

Fig. 15.1. Scatter diagram of fastener torque values before and after vibration tests.

When an analyst uses a scatter diagram, he or she must be careful about making a positive statement about the relationship, even if the variable points have a narrow scatter. The variables being studied may indicate a relationship but may be completely unrelated to each other. An example would be the association between the football Superbowl scores and what happens to the stock market during the year. Some analysts have attempted to fabricate credence to this relationship, but it should be obvious that no relationship exists.

Simple scatter diagrams can be examined further by performing a correlation and regression analysis on the data. The analysis results in correlation and regression coefficients that introduce greater in-depth knowledge about the data. The references at the end of the chapter provide further directions and meaning to scatter diagrams and to the analysis.

Histogram

Another simple tool used to analyze data and to improve quality or reliability is the histogram. Figure 15.2 shows a typical bar-graph relationship between measured shaft diameters and the frequency of occurrence at each dimension. The histogram also shows the drawing limits of the diameter and how closely the measured values are within the tolerance.

The example shows a simple and clear picture of how the data is slightly to the right of the specification mean, how the dimensions are spread out, and how the histogram is shaped. The picture is descriptive and discrete on the relationship of the data points with the dimensions and their tolerances. Further sophistication can determine the estimated values that will be above and below specification, but the simplicity of this chapter does not delve into this technique. The references supply details on how this can be accomplished.

Figure 15.3 shows some typical patterns of different histograms. These originally were introduced in 1946, but the distribution patterns tell much about what is happening today. A brief explanation is provided with each histogram to simplify the interpretation of each chart.

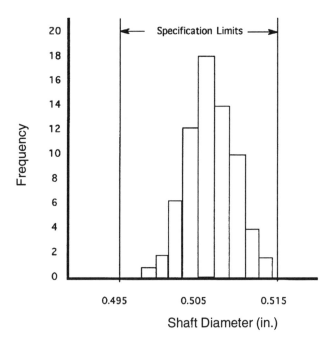

Fig. 15.2. Histogram of shaft diameters.

Certain cautions must be used when interpreting histograms:

1. Be aware that the sample represents the population as well as it can but is only a sample.

2. Be sure to understand and relate what the histogram represents (e.g., date of data collection, different shifts or days or lots).

3. Do not make rash conclusions based on a small sample size. Some references state that a sample of 25 is an absolute minimum, but even more important is observation of how the data is spread and the ability to make some confident statements about the data.

4. Remember that a histogram is a sample of the whole. Be careful of any statements about what it represents.

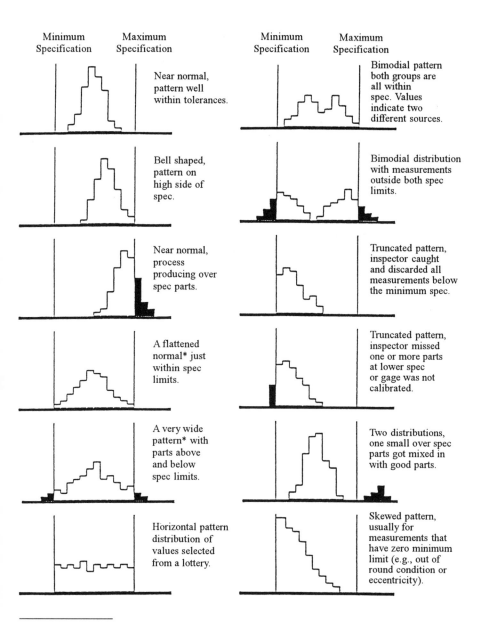

Minimum Specification	Maximum Specification		Minimum Specification	Maximum Specification	

Near normal, pattern well within tolerances.

Bimodial pattern both groups are all within spec. Values indicate two different sources.

Bell shaped, pattern on high side of spec.

Bimodial distribution with measurements outside both spec limits.

Near normal, process producing over spec parts.

Truncated pattern, inspector caught and discarded all measurements below the minimum spec.

A flattened normal* just within spec limits.

Truncated pattern, inspector missed one or more parts at lower spec or gage was not calibrated.

A very wide pattern* with parts above and below spec limits.

Two distributions, one small over spec parts got mixed in with good parts.

Horizontal pattern distribution of values selected from a lottery.

Skewed pattern, usually for measurements that have zero minimum limit (e.g., out of round condition or eccentricity).

---

* Distribution is with positive kurtosis. For a normal distribution the kurtosis would equal zero.
** Positive skewness-pattern tails to the right. For a normal distribution skewness equals zero.

Fig. 15.3 Histogram patterns and their interpretations. (From Quality Progress Magazine, *September 1946)*

Flow Chart

One quality tool that has been in use for many years is the flow chart, which is used to demonstrate the movement of a process, paperwork flow, or machining sequence. The flow chart presents a pictorial view of the process and shows the relationship between the various elements that make up the process. Figure 15.4 shows the process flow of a Failure Reporting, Analysis, and Corrective Action System (FRACAS). It is identical to the flow chart depicted in Fig. 8.2. The chart shows where and how reports are initiated, to whom they are sent, and what happens to the reports after they are received. Everyone in the pattern of events knows for what he or she is responsible and for what other individuals are responsible. The flow chart can be made in almost any manner or style. Drawing symbols, pictures, and engineering nomenclature are typical examples (e.g., rectangles, circles, squares, and triangles). The important factor is how the overall picture shows the process flow and explains in some manner the "who, what, when, and where" of the process items as they relate to each other.

In constructing a flow chart, several factors should be considered, as follows:

1. The person(s) making the chart should be knowledgeable about the process.

2. If possible, the flow process should be seen on one page as an entity. If the flow chart extends to separate pages, the pages should be taped together or arranged in numerical sequence, identifying the last block with the same number to start on the next page.

3. Compare the first draft of the flow chart with the actual process, and look for improvements and better ways to accommodate the flow.

4. Continually ask questions about what is happening exactly, and attempt to improve the process wherever possible.

The end result of a flow diagram should be a chart that clearly shows how a process works, who is responsible for specific sections, who does what, and when and where certain items should be done. The chart should reflect the most efficient flow and "cover all bases" of the process. It should be reviewed periodically to determine where continuous improvements can be made.

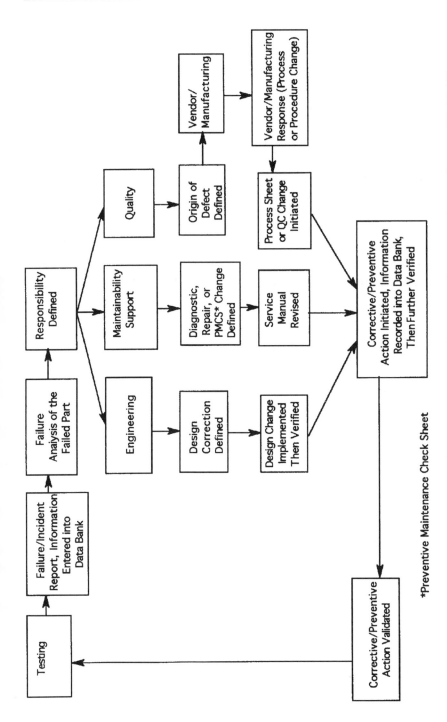

Fig. 15.4. Flow diagram for a Failure Reporting, Analysis, and Corrective Action System (FRACAS).

*Preventive Maintenance Check Sheet

Cause and Effect Diagram

The cause and effect diagram is known by other names. These include the Ishikawa diagram, named after the person who developed the technique; the fishbone diagram, because the diagram resembled the skeleton of a fish; and the feather diagram, because it was viewed as a feather. Figure 15.5 shows a typical cause and effect diagram of a vehicle brake problem. The purpose of the diagram was to relate causes and effects. Generally, the problem as defined is indicated in the "head block" (effect) of the fishbone, and all items that can be related to the problem (causes) are indicated in the bone structure of the diagram. These usually relate to (but are not limited to) management, machines, material, money, manpower, methods, measurement, and environment—known as the main causes (also known as the "Seven M's" plus the environment)—and then to the related subcauses. The Seven M's originally were considered in the analysis, where the non-applicable ones were deleted.

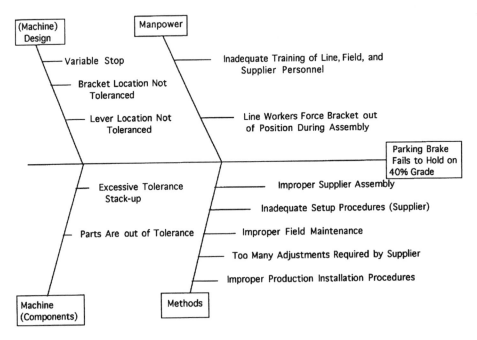

Fig. 15.5. Cause and effect diagram of a brake problem.

Brainstorming is explained later in this chapter. However, it is a technique of collective group thinking and discussion which provides a means of finding solutions to problems. It can be used to determine the possible causes of the problems. These causes usually fall under the Seven M's and environment. After the problem is identified clearly and the causes and subcauses are stated as well as possible, a cross-functional team can determine which causes are most responsible for the problem and then take corrective action as necessary. See the following section on the Pareto chart.

After the cause and effects diagram is constructed, the analyst or team can review its contents to determine where to work out possible solutions to the problem. This usually is done best as a team approach because the various specialists and skill levels can contribute to the solution.

Pareto Chart

The Pareto chart is a form of a histogram that helps determine which problems have the highest order of occurrence. The idea was developed in the nineteenth century by an Italian economist named Vilfredo Pareto (1848–1923) while he was attempting to find the distribution of wealth. Pareto discovered that a large share of the wealth was shared by relatively few people. Joseph Juran (b. 1900), in developing his own theories, discovered that quality problems related in much the same way. Thus, the term Pareto principle developed.

In effect, the Pareto principle tells us that roughly 80% of the problems are caused by 20% of the possible sources to the problem. Figure 15.6 shows an example of the weighing of the causes related to the brake problem in the cause and effect diagram shown in Fig. 15.5. Note that the design complexity, variable stop, and bracket angles that do not have a tolerance showed the highest weights, which helped determine on what the team should concentrate as possible solutions to the problem.

Data collection can be from almost any source, and the exact 80/20 principle will not always be true. Quantitative data can be compared for different shifts, machines, materials, operations, reasons for scrap, warranty claims, reasons for error, material handling, or any imaginable characteristic. Generally, the most frequent cause is placed on the left side of the chart, and other causes are

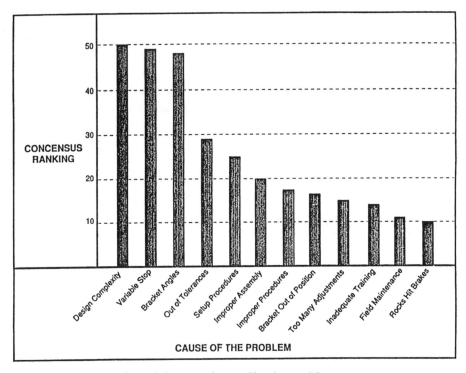

Fig. 15.6. A Pareto chart of brake problem causes.

placed in the descending frequencies to the right. The picture becomes very clear as to which causes are the most frequent and which ones are the least frequent. As the corrective actions are taken on the highest-frequency causes, their frequencies reduce. The chart is updated again to find the 80% problems that are caused by the 20% sources. Amazingly, the pattern of the chart is repeated. This can be continued until, at least theoretically, all problems are eliminated.

Control Chart

The control chart has been used since the 1920s when Walter Shewhart (1891–1967), first developed it for Bell Laboratories. W. Edwards Deming (1900–1993) incidentally worked with Shewhart during this time. Basically, the control chart is a running chart with a statistically determined average ($\overline{\overline{X}}$) and upper and lower control limits ($\overline{\overline{X}} \pm A2\overline{R}$) for process dimensions or other measurable characteristics. The A2 factor is determined from a table of values (see references by Duncan, Grant, Juran, or any quality book that has the table). The limits of the chart are determined by operating a process and measuring the variability of the product produced by the process. The points usually are sample averages with ranges of the characteristic being measured. Figure 15.7 shows a typical $\bar{X}$ and R control chart. The control limits usually are set at the average $\pm 3\sigma$'s, which are derived from the equation ($\bar{\bar{X}} \pm A_2\bar{R}$) after a continuous run is made, without making any adjustments to the process.

Other types of control charts—such as the c, p, np, and u charts, also can be used to analyze data. The charts, respectively, are for number of nonconformities (c), fraction or percent of nonconformities (p), number nonconforming (np), and defects or nonconformities per unit (u). References at the end of this book provide additional information on how $\bar{X}$ and R charts and other control charts are constructed and utilized. Control charts frequently are used in production processes, but they also are a powerful tool for R&M and analytical purposes.

Check Sheet

A check sheet is a logical technique used to tally how frequently specific events occur. In effect, the check sheet describes "what is going on." It usually separates fact from fiction and provides information and answers about a particular product or process. Figure 15.8 shows a typical check sheet that describes one set of ball joint torque readings from the various wheels on a truck, before and after a test run. The charts opened the door to many questions as to why the torque readings for the wheels were different among the wheels and why such a large difference was found among readings before and after the test run. Sometimes, a visual analysis of the check sheet can provide the answers.

Check sheets come in almost any form for collecting useful information. They can be drawn freehand, but they should include criteria such as the following:

1. What is being looked for and what question(s) must be answered

2. The best way to answer the question(s)

171

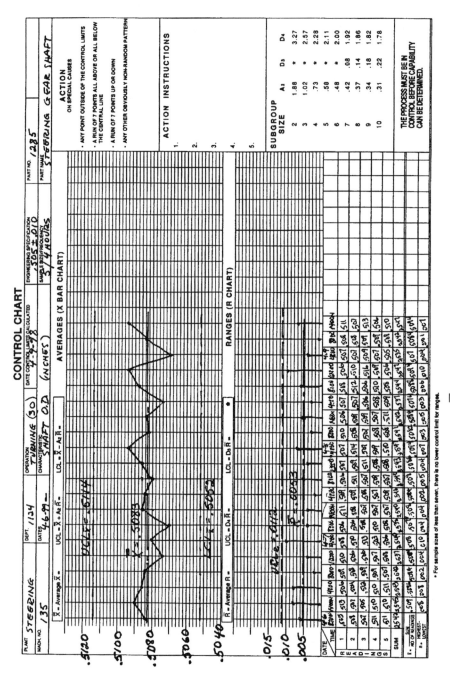

Fig. 15.7. $\overline{X}$ *and R chart of shaft diameters.*

172

3. When the data should be collected and for how long

4. A form design that is easily used, relatively simple, and understandable

5. Data collection with minimum effort, chance of error, and the least interference to the process

6. Recognition of the importance, accuracy, and usefulness of what is being collected

If check sheets are formulated and developed in a dependable and reliable manner, and if data is gathered accurately, they can be useful in the quality reliability improvement process.

Other techniques that are useful in the quality improvement process are explained here in brief detail. These are presented in this chapter to demonstrate additional techniques available to the analyst.

Brainstorming

This was mentioned previously in this chapter. In simplest terms, brainstorming is the process by which collective ideas are gathered concerning solutions to a particular problem. At initial onset, the ideas should flow freely and be expressed, however ridiculous they may seem. Generally, the ideas are listed on a blackboard or easel pad, allowing everyone to see the ideas and to generate new ideas. Later, when the generation of ideas is exhausted, the ideas themselves can be analyzed in detail.

Stratification

Stratification is the assurance of gathering samples that most accurately describe a population. In other words, subgroups of a population should be represented proportionally in samples taken from the population. An example of stratification would be if someone wanted to analyze the reliabilities of machinery from several plants or departments and could examine only samples from each. The analyst would sample a proportionate number from each of the plants, ensuring that the representative sample sizes are in like percentages as in each location.

Data Identification *BALL JOINT TORQUE READINGS ON THE LF WHEEL, NO. 1 LOCKNUT.*

Preparer

Date 5/5/98

BEFORE TEST RUN

Tally	Cell Mid Point	Freq.	Accum. Freq.	% Under
	24	0		
HHT III	23	5	5	2.7
HHT-HH HH HHT HH HH HH HH HH HH HH HH HH HH HH	22	73	78	41.7
HH HH HH HH HH I HH HH HH HH HH	21	51	129	69.0
HH HH HH HHT HH II HH HH HH HH HH	20	52	181	96.8
III	19	3	184	98.4
I	18	1	185	98.9
	17	0	185	98.9
	16	0	185	98.9
I	15	1	186	99.5
	14	0	186	99.5
I	13	1	187	100.0
	12	0		
TOTAL				

AFTER TEST RUN

Tally	Cell Mid Point	Freq.	Accum. Freq.	% Under
	24			
	23			
	22	0		
III	21	3	3	1.6
HH I	20	6	9	4.8
HH HH III	19	13	22	11.8
HH HH HH HH II HH HH HH	18	33	55	29.4
HH HH HH HH HH	17	25	80	42.8
HH HH HH HH IIII HH HH HH	16	34	114	61.0
HH HH HH I HH HH	15	21	135	72.2
HH HH HH HH	14	19	154	82.4
HH HH HH HH HH HH III	13	28	182	97.3
IIII	12	4	186	99.5
I	11	1	187	100.0

Fig. 15.8. Check sheet for ball joint fastener torques.

Run Charts or Trend Charts

These are other types of charts for representing data. They relate closely to control charts but usually do not show control limits. They are used primarily to monitor a process or demonstrate a trend to determine what is occurring to a long-range average. They are constructed easily and can be used by simply following a time sequence:

- What is happening to the production of a machine
- Scrap that is being produced
- Number of complaints from a customer(s)
- Number of errors in an operation
- Any measure of overtime or sequence
- Tool wear on a machine

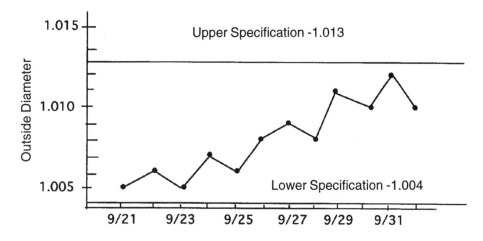

Fig. 15.9. A run chart showing tool wear influence on an outside diameter.

Typically, a run chart appears similar to the graph shown in Figure 15.9.

One danger in using a run chart is that analysts tend to project into the future, believing that the run or trend will continue. They are likely to neglect the natural variation of data, putting overemphasis on the fluctuation of the points. The chart is especially useful in detecting shifts in the average or changes in the trend. However, care should be exercised to ensure that random variation is a natural occurrence. Generally, nine continuous points on one side of the average, or six points steadily increasing or decreasing, indicates a possible shift in process that at least should be investigated.

Nominal Grouping Technique

The Nominal Grouping Technique (NGT) is simply a matter of organizing the data of a particular problem so that all of the information can be considered and prioritized to some extent. It is a method of ranking by consensus. It allows every team member to rate issues without being pressured by the other members. The method can concern problems, solutions, importance of issues, or decisions on projects. This technique more or less neutralizes the loudest concern or one that is endorsed by a highest authority. Everyone in the group has an equal say concerning the issue in question. The technique first lists the problems. Then each member of a team votes, weighing the issues to determine which ones they believe are the most important. (These issues can be numbered from one to ten or some other weighing scheme.) The values are then averages to determine which ones should receive first consideration.

The example shown on the Pareto chart (Fig. 15.6) illustrates what was first developed by a brainstorming session. Then, with NGT, the causes were ranked by consensus of the problem-solving team. The intermediate result was the Pareto chart that allowed the team to put its resources into what the team thought were the most pronounced causes. The end result was a solution to a complicated problem.

Force Field Analysis

In this technique, which was developed by Kurt Lewis, the driving forces are viewed against the restraining forces. In other words, an analyst would list all the driving forces that relate to a concern and then correspondingly list the restraining forces that would hinder or not permit the driving forces to happen. The driving forces can be developed from a brainstorming session or from the NGT, and then analyzed by a team or by an individual.

Process Capability

Process capability relates closely to control charts. It is the built-in consistency of a process that is reflected in the product that is produced. Control charts classically are used to determine process capability which, in turn, determines whether the process with its natural variation meets the tolerance limits of the product. However, remember that control charts also are used for analytical purposes to determine differences in machine performance, operations, or almost any factors that make up a process.

Gantt Chart

This type of chart is a listing of activities or tasks that usually relate with time. Each task is listed in a left-hand vertical column, accompanied by a horizontal bar that extends from a time beginning period for each task to a time slot that represents a scheduled completion date. The chart is an excellent schedule-monitoring tool. It is extremely useful for organizing task assignments while allowing responsible personnel or departments to know the status of their activities as related to other tasks. The status of all activities can be viewed simultaneously. Modifications regarding action items and assignees, schedules and completion indicators, key dates and activities, the relationship of the tasks, and other additions also can be utilized in the chart. Figure 15.10 shows a Gantt chart that depicts the status of the progress that is being made on a QS-9000 program.

Plan, Do, Check, Act (PDCA) Cycle

No text on R&M analysis and instructions can justifiably omit the use of Shewhart's or Deming's Plan, Do, Check, Act (PDCA) cycle when considering problem solving. The chart sometimes is referred to as the Plan, Do, Study, Act (PDSA) cycle. The *Process Improvement Guide* uses the PDSA terminology, and it also contains additional information on many of the other problem-solving techniques. The PDCA method also is known by other terms, but most of them are basically the same process. The following steps make up the PDCA cycle. Figure 15.11 presents the typical format of what Shewhart thought and what Deming derived.

1. **Plan:** Formulate for what the team is responsible, know its mission, determine what is required to accomplish the task(s), outline the testing and observations, and decide on a schedule.

2. **Do:** Accomplish what the team has planned or changes that have been requested or desired, preferably on a small scale at first.

3. **Check:** Observe and note the accomplishments and the effects of the change action. Be sure to maintain good records.

QS-9000 Tasks	Months - 1998-2000																	
	Oct	Nov	Dec	Jan	Feb	Mar	Apr	May	Jun	Jul	Aug	Sep	Oct	Nov	Dec	Jan	Feb	Mar
Rough Documentation Developed (1)	•																	
Have Gap Analysis Conducted (2)		—•																
Review Findings and Observations		—•																
Initiate and Implement CA/PA			—•															
Train All Employees			—•															
Select and Train Internal Auditors				—•														
Bring Level I Doc's to Compliance Standard (3)				----•														
Bring Level II Doc's to Compliance Standard (3)					----•													
Bring Level III Doc's to Compliance Standard (3)					-------	---												
Develop Forms to Compliance							---------											
Ensure Implementation of Procedures								—•										
Perform Internal Audits									—									
Initiate and Implement CA/PA										—								
Prepare Pre-audit Readiness											—							
Pre-registration Audit by Registrar												—						
Initiate and Implement CA/PA														—				
Certificate Audit by Registrar															—			
Initiate and Implement CA/PA																-------		
Obtain Certification																		--

(1) Management Rep selected and rough documentation developed by this point in time.
(2) Can either be by a self analysis or by an independent consultant.
(3) This includes the writing, implementation, and compliance or all the QS-9000 elements.
—, Time scheduled or expended
• Task completed

Fig. 15.10. Gantt chart for XYZ QS-9000 progress and status.

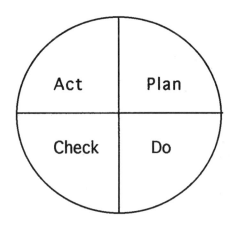

Fig. 15.11. The classic PDCA diagram.

4. **Act:** Study the results and determine if they are what the team was trying to accomplish. Determine what was learned and how that learning can contribute to the team project, or if any adjustments or changes must be made. If the results do not meet the expected requirements, return to Step 1 and plan again, using whatever information was derived from the previous analysis.

5. Repeat the cycle as necessary until the team reaches the result of its mission.

In addition to the quality tools, other tools are available to R&M analysts and cross-functional teams. These are the seven management tools that also can be used in the analysis of information. The seven management tools originated in Japan around 1972 and have since been modified in the United States while remaining quite similar. Goal/QPC (see *The Memory Jogger II* reference) was heavily involved in the modification process, which resulted in the seven management and planning tools listed here. These are explained briefly in the subsequent sections.

1. Affinity diagram
2. Tree diagram
3. Process Decision Program Chart (PDPC)
4. Matrix diagram

5. Interrelationship digraph
6. Priority matrices
7. Activity network diagram

Affinity Diagram

This is a collection of many ideas, opinions, or facts separated into smaller groups that relate to each othc.. In effect, the technique separates the information and then brings it all together into similar thought patterns or like descriptions. After the information is together in various groupings, the team can analyze the data more systematically. Figure 15.12 displays the basic tasks required for implementing a QS-9000 program. The chart separates the tasks into four main categories of related items. The diagram also could include personnel responsible for the tasks and expected completion dates; however, a Gantt chart is a more efficient method of overall scheduling.

Tree Diagram

This is another systematic method similar to an organization chart or a family tree that is driven by an overall goal or objective. Then by levels, ideas are generated concerning how a team would achieve that goal, followed by each succeeding level until a final level is reached. Then the process is continued until the team believes it has included all levels of results it could possibly attain to solve the stated problem or goal. Surprisingly, as the team thinks it has finalized its thought process by confirming that it has included all the means that would lead to a successful objective, other ideas usually come to mind. These also should be included as the team reviews all the inputs and draws some conclusions on how it can reach its goals. Figure 15.13 depicts a scrap reduction program that includes the undertakings that would help accomplish such a goal. As the program is being implemented, other related tasks can be added to the diagram to make the program more effective.

Process Decision Program Chart

The Process Decision Program Chart (PDPC) shows plots of decisions that a team believes are necessary from the start of the goal or project to its final completion. All possible events and contingencies are included in the projected decision making. As

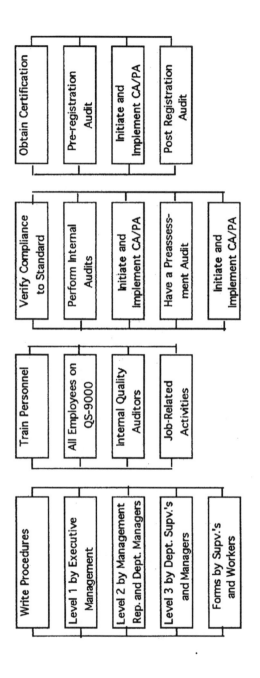

Fig. 15.12. Affinity diagram for implementing QS-9000.

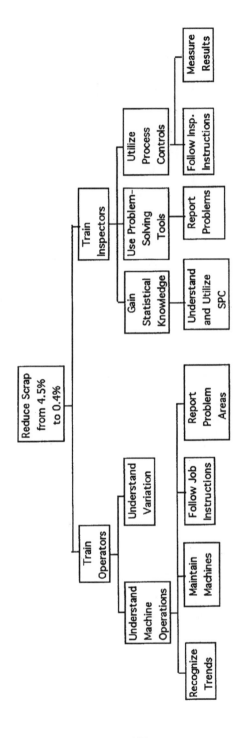

Fig. 15.13. Tree diagram for reducing scrap from 4.5% to 0.4%.

in a tree diagram, contingencies and possibilities are sequenced together by connecting similar ideas, running them all in the direction to the final goal. Figure 15.14 is an extension of a tree diagram because it includes contingencies for the process if the goal appears to be faltering.

Matrix Diagram

A matrix diagram often is done unconsciously when individuals want to compare one set of information to another set of data. Personnel are matched against training records. Departments and other individuals are compared against responsibilities. Components are equated with the functions they perform. The comparisons are endless regarding what can be done with matrices and the various formats they can take. The idea is to match the information in such a manner that it will be most beneficial to the analyzing team. Figure 15.15 represents a matrix diagram that can be used in preparation for a QS-9000 quality audit. The chart shows the 20 elements of QS-9000 and space for the names of the auditors and auditees who are expected to be interviewed. If appropriate, the scheduled meeting times for the interviews also are included, but auditors should remain flexible.

Interrelationship Digraph or Diagram

This is a circular grouping of activities or issues, but they are connected together by way of their interrelationships. The technique allows a team to systematically identify, analyze, and classify the relationships that exist among the issues. It spurs team members to think of multiple issues and how those issues relate with one another. Each activity is laid out individually, and a question is asked between each pair of activities: "Is there a cause/influence (C/I) relationship between these two activities?" If the answer is "yes," a line is connected between the two issues that are being considered, and an arrow is pointed from the direction in which the C/I is stronger. If no relationship can be found, no line is drawn. After all the pairs are appraised, the arrows in and out are counted for each activity and so indicated on each issue. By identifying the ins and outs for each activity, it is possible to identify the key drivers (i.e., the greatest number of outgoing arrows) and the key outcomes (i.e., the greatest number of incoming arrows). The key drivers are the activities that generally are the root cause(s) of the issue being analyzed. The activity that has the most arrows going out usually is the one on which the team concentrates

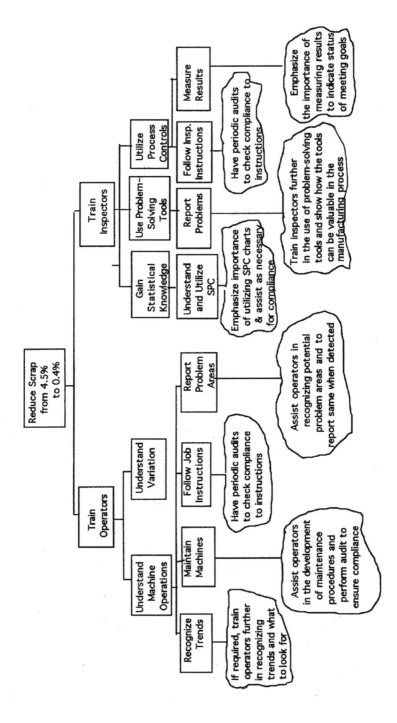

Fig. 15.14. Tree diagram for reducing scrap from 4.5% to 0.4%, extended to a PDPC chart.

	Auditor/Auditee (1)									
				(2)						
QS-9000 Elements										
Management Responsibility										
Quality System										
Contract Review										
Design Control										
Document and Data Control										
Purchasing										
Control of Customer-Supplied Product										
Product Identification and Traceability										
Process Control										
Inspection and Testing										
Control of Insp., Measuring, and Test Equip.										
Inspection and Test Status										
Control of Nonconforming Product										
Corrective and Preventive Action										
Handling, Stor., Pack., Preserv., and Delivery										
Control of Quality Records										
Internal Quality Audits										
Training										
Servicing										
Statistical Techniques										
Customer-Specific Requirements										

(1) List auditors assigned to auditees.

(2) Names of individuals with scheduled meeting times.

◯ - Coordination of auditor/auditee and QS-9000 element.

⊙ - No nonconformance found as audit is performed.

⊘ - Nonconformance(s) found

Fig. 15.15. Matrix chart for auditor/auditee assignees.

185

first. For planning purposes, the team focuses on the key outcome activities. In our example, Fig. 15.16 portrays the interrelationship among the charges that relate from the time a customer wishes to contract a manufacturer to make or design a machine to the actual production of the reliable product the customer desires. The chart shows the mishmash of lines. However, on close examination, it can be seen that project management (or the support of management) is the key driver and that the manufacture of a reliable product is the outcome.

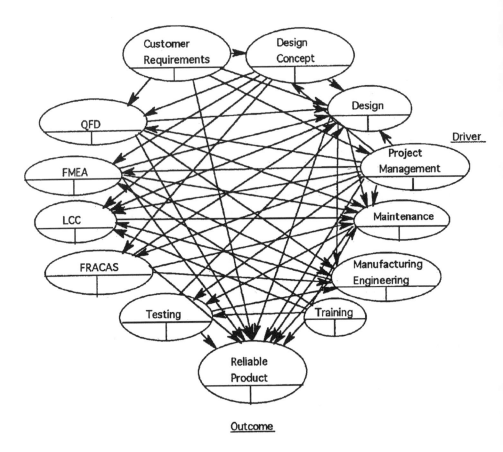

Fig. 15.16. Interrelationship digraph of concerns in producing a reliable product.

Priority Matrices

This is a broad extension of the matrix diagram because it utilizes several matrices in its analytical approach. Its strong points include consensus criteria and a combination of interrelationship digraph and matrix methods. To use the technique, key issues and concerns first must be identified. Then alternatives can be generated and determination can be made regarding what options should be used. The technique can become complicated because it may include too many items. However, the idea is to lay out the information in matrices and then prioritize all the options. *The Memory Jogger II* provides an in-depth procedure on how the priority matrices works.

Activity Network Diagram

This diagram is similar to, but slightly more complicated than, the affinity diagram (which sometimes is called the arrow diagram). It includes the activities, milestones, and critical times that make up a problem-solving task or project, and how they are related and tied together. The activities are connected jointly by the sequence of events, similarly as in a Concurrent or Simultaneous Engineering, or a Program Evaluation and Review Technique (PERT) chart, but also including the options within the grouping, together with earliest and latest start and finish times for each activity. After completion of the diagram, the team can review and analyze the chart to determine where improvements can be made.

Figure 15.17 shows an activity network of a design project. The chart presents a rough sequence of the charges and the assignments involved in the process of designing a machine. The estimated timing can be added to each responsibility and then summed in their line arrangement. The timing is not shown on the chart; however, testing and analysis definitely is the critical path of the network (or the sequence of events that take the longest time). The other tasks should be related to this path. When the network chart is established, management is responsible for ensuring that the obligations are completed as scheduled, that the customer requirements are met, and that the necessary resources are supplied to accomplish the job.

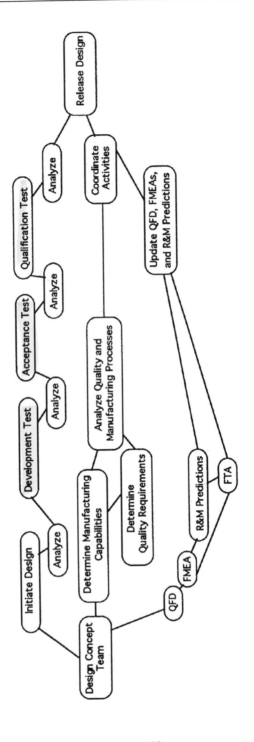

Fig. 15.17. Activity network diagram for a design project.

Summary

The intent of this chapter was to present an outline of the quality and management tools available to assist in the never-ending race to initiate continuous quality improvement and to enhance reliability. The tools can be utilized individually or in combinations with each other. Much depends on what is trying to be accomplished and how effective the tool will be for the particular problem or project. In summary, the design engineer must recognize the task at hand. If problems develop during the project, those problems must be identified and defined, analyzed for root cause, and then evaluated and prioritized. The final responsibility is to initiate and implement the corrective and preventive action, which then is verified.

Use of these techniques should improve responsiveness to customer needs, produce better-quality parts, reduce costs, and encourage proactive ideas for more effective and efficient processing. The chapter presented some of the more popular and known methods in this regard. The more in-depth details of the techniques can be found in the references presented throughout this book.

References

Bajaria, Hans J., and Copp, Richard P., *Statistical Problem Solving*, Multiface Publishing Co., Garden City, MI, Amsterdam, 1991.

Brassard, Michael, and Ritter, Diane, *The Memory Jogger II, A Pocket Guide of Tools for Continuous Improvement and Effective Planning*, Goal/QPC, Methuen, MA, 1994.

Duncan, Acheson J., *Quality Control and Industrial Statistics*, 3rd ed., Richard D. Irwin, Homewood, IL, 1965.

Grant, E.L., and Leavenworth, R.S., *Statistical Quality Control*, 4th ed., McGraw-Hill, New York, 1989.

Holmes, Susan, and Ballance, Judy, *Process Improvement Guide*, 2nd ed., Air University-Maxwell Air Force Base, AL, 1994.

Juran, Joseph M., *Juran's Quality Handbook*, 5th ed., McGraw-Hill, New York, 1999.

Juran, J.M., and Gryna, Frank M., *Quality Planning and Analysis*, 3rd ed., McGraw-Hill, New York, 1993.

Periodicals:

Burr, John T., "The Tools of Quality—Part I: Going with the Flow (Chart)," *Quality Press*, June 1990, pp. 64–67.

Burr, John T., "The Tools of Quality—Part VI: Pareto Charts," *Quality Press*, November 1990, pp. 59–61.

Burr, John T., "The Tools of Quality—Part VII: Scatter Diagrams," *Quality Press*, December 1990, pp. 87–89.

Juran Institute, "The Tools of Quality—Part IV: Histograms," *Quality Press*, September 1990, pp. 75–78.

Juran Institute, "The Tools of Quality—Part V: Check Sheets," *Quality Press*, October 1990, pp. 51–56.

Sarazen, J. Stephen, "The Tools of Quality—Part II: Cause and Effect Diagrams," *Quality Press*, July 1990, pp. 59–62.

Shainin, Peter D., "The Tools of Quality—Part III: Control Chart," *Quality Press*, August 1990, pp. 79–82.

Chapter 16

Safety

Safety is another item the design engineer should consider. Fortunately, he or she has help in this area similarly as in others. Safety goes hand in hand with R&M and with tooling and equipment. The chief concern for the designer is that when a machine has a failure, the resultant anomaly causes no serious accident or injury. Safety engineers have established that machining environments pose a greater number of risks than many manufacturing situations. Additionally, the risks usually are more complicated. Therefore, design engineers should keep this in mind when designing tooling and equipment. There are numerous safety standards, regulations, authorities, and agencies. The designer must know and understand these, but he or she also must comprehend how to interlock safety into the product. He or she should have a calculated knowledge of what the safety margins are and how they must be applied.

No compromise can be made about safe operation of equipment, as well as maintenance without risk to operators, maintenance personnel, and anyone involved with the machinery. No matter how good the design is for a product, it is unacceptable if the product cannot be operated and maintained safely.

One of the safety guidance determinants is the safety factor (SF). Several R&M authors present safety margins and safety factors in different ways, but this should not deter the design engineer on how safety should be regarded. This chapter proposes that the measure of safety is the amount of spread or buffer between stress applied to a product and the ability of the component to operate under the applied stress. This circumstance also can relate to the load-carrying capacity of a component or its strength versus the loads applied to a part or the system. Figure 16.1 pictorially shows this relationship.

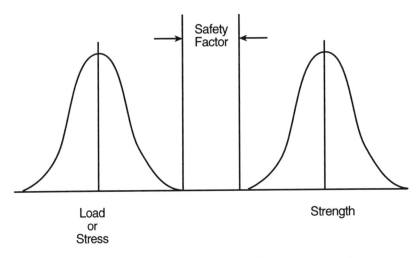

Fig. 16.1. Relationship between load and strength.

Distribution should be known in order to calculate the safety factor, the means, and standard deviations of the load and capacity.

The equation for determining the safety factor in the relationship shown in Fig. 16.1 is as follows:

$$SF = \frac{\bar{S} - \bar{L}}{\left(\sigma_S{}^2 - \sigma_L{}^2\right)^{\frac{1}{2}}}$$

then

$R = \phi(SF)$ Looking up the probability of SF in the Normal Distribution Table (found in most textbooks on statistics or quality assurance)

where

$$SF = \text{Safety factor}$$

$\bar{S}$ = Mean of the strength

$\bar{L}$ = Mean of the load (or stress)

σ_S = Standard deviation of the strength

σ_L = Standard deviation of the load

R = Reliability between the strength and the load (or probability of the load being greater than the strength)

$\phi(SF)$ = Probability of the load being greater than the strength, using the Normal Distribution Table

It readily should be seen from the diagram that the farther apart the load distribution and strength distribution are, the higher the reliability or the safety factor.

If the standard deviations of the two distributions are unknown or cannot be estimated, another method of determining the safety factor would be to apply the following equation:

$$SF = \left[\frac{NR - AL}{AL} \right] 100\%$$

where

SF = Safety factor
NR = Component net rating
AL = Applied load

This equation demonstrates that the safety factor is the nominal margin of safety separating the net rating estimate and the applied load or stress. In this case, no consideration is given to the variability of these estimators.

Juran and Gryna take another approach to the safety factor and safety margin. In their Quality Planning and Analysis, they state that the safety factor is the ratio of average strength to the worst stress expected. Figure 16.2 shows their approach to calculating the safety margin. They expand the definition of the safety factor to what they define as the safety margin, or the difference between average strength and the worst stress divided by the standard deviation of the strength distribution (note that the safety factor they define does not consider the distribution variance of the stress) or

$$\text{Safety Margin} = \frac{\text{Average Strength} - \text{Worst Stress}}{\text{Standard Deviation of the Strength}}$$

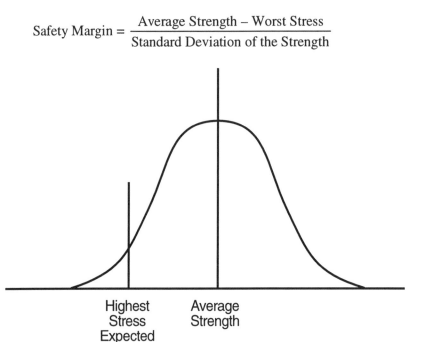

Fig. 16.2. Relationship between average strength and worst stress.

Then, using the number of standard deviations the average strength is above the worst stress, the reliability of that characteristic can be found by referring to the Normal Distribution Table as was accomplished in the Fig. 16.1 association. The approach is conservative because the relationship does not take into account the variation in the stress as in Fig. 16.1. The relationship is shown diagrammatically in Fig. 16.2.

The design engineer should not be confused with these different descriptions of what is considered a safe design. The important determinant is that he or she understands the probability (or reliability) of a stress or load on a component being higher than the strength of the part. With this in mind, the former calculation as shown in Fig. 16.1 is a good measure of safety. This should be at a level that is high enough to what the user or the safety engineer defines.

The safety factor or reliability initially is set by the user, safety engineer, or management of the manufacturer. The safety factor can vary, depending on the failure history of the part or similar part, different applications and environments, or the structural ruggedness of the components. Usually, minimum and maximum safety factors are required, allowing some flexibility in the design. You could argue that there should be no tradeoffs with regard to safety; however, if the designer designs the product with a safety factor far above the required maximum, the cost of the product increases significantly.

Risk factors are the recording of potential hazards as they can be identified, and then analyzing and disposing of them to reduce the overall danger in a product. The safety engineer reviews the design to determine the safety risks in the prototype. He or she prepares a list that is used to reduce the risk of the components involved. This list then is kept up to date because testing also reveals safety concerns as they appear. Generally, it is a good idea to prioritize the risk factors in the order of what the safety engineer believes to be the most serious problems. He or she focuses on these first, and then works on the other concerns.

From a design point of view, what can the design engineer do to enhance safety in machinery? There are many methods, including the following:

- Hazard analysis

- Safety interlock

- Safety notes in the manuals

- Use of diagnostics, "breakpoints," low-stress components, and protective arrangements

- Other safety devices

These methods are discussed in detail in the following paragraphs.

Hazard Analysis

Hazard analysis is the review and correction of a real or potential condition that can cause unintentional injury or death to humans, or damage to or loss of equipment or property. Hazard analysis should not be confused with hazard rate. Hazard analysis concerns the correction of conditions where hazard rate relates to the instantaneous failure rate. In this chapter, we are concerned with all conditions that relate to safety. The hazards are classified in four levels:

1. Negligible or minor
2. Marginal
3. Critical
4. Catastrophic

As in the risk factors, the priority of correction starts with the potential catastrophic hazards and then proceeds to the minor ones. Depending on the seriousness of the safety concern, tradeoffs may be compromised, although all safety concerns should be addressed.

In some sense, hazard analysis is similar to an FMEA because the safety engineer (or reliability engineer) tries to determine all hazards that possibly could occur. He or she uses a thorough and logical analysis of all potential risks and then applies corrective actions, in the same manner as in an FMEA. Experience and forward thinking are tremendous assets in hazard analysis, which can be lengthy, costly, and overtaxing. However, there is a payoff.

Safety analysis can be inclusive to a failed parts analysis as a by-product. As a failed parts analysis is performed, the analyst informs the safety engineer to determine the hazard risk considered for a redesign. The safety engineer assesses characteristics such as safety interlock, safety margins, personal and property safety, or any other safety issues. He or she discusses the concern with the designer, who then makes appropriate revisions as necessary.

Safety Interlock

Safety interlock is a design feature that prevents access to a service envelope while a hazardous potential exists within it. The design plan is to render the hazard harmless when the service envelope is accessed. A switch built into an access cover is a good example of an "interlock." If the access cover is removed, the switch automatically becomes electrically open. The interlock feature is an important essential, especially if untrained personnel have an opportunity to work on the equipment.

Environmental factors are a concern to the safety engineer. Different environments must be considered. These include, but are not limited to, the following:

- Unpredictable action of people (i.e., if someone can use an item the wrong way, the item will be used in its unintended way)

- Biological organisms

- Contaminants

- Electrical fields and interference

- Electromagnetic forces

- Electrostatic discharges

- Explosive potentials

- Operating characteristics of nearby equipment

- Abnormal pressures

- Storage facilities

- Extreme temperature variability

- Transportation

Any one of these factors can play havoc with equipment and potentially may do harm if the factor is not considered. Both safety engineers and designers should consider these factors. A logical place to consider these factors would be when FMEAs and FTAs are developed. These methods are discussed in detail in Chapter 5.

Industrial and government standards and regulations must be reviewed and followed when applicable. The safety engineer should be knowledgeable about the standards that apply to his or her equipment and know what tests must be performed. The government safety standard MIL-STD-882, System Safety Program Requirements, provides a uniform guidance for establishing and implementing a safety program. It usually applies to Department of Defense (DOD) contracts, but it is a good guidance document for any safety program. Distinct safety-related tasks are specified in the standard that can be required in a contract. The following tasks are included in the standard:

Task No.	Task Description
	100 Series Tasks
100	System Safety Program
101	System Safety Program Plan
102	Integration of Subcontractors
103	Program Reviews
104	System Safety Working Group
105	Hazard Tracking
106	Test and Evaluation Safety
107	Progress Summary
108	Key Personnel Qualifications
	200 Series Tasks
201	Preliminary Hazard List
202	Preliminary Hazard Analysis
203	Subsystem Hazard Analysis
204	System Hazard Analysis
205	Operation Hazard Analysis
206	Health Hazard Assessment
207	Safety Verification
208	Training
209	Safety Assessment
210	Safety Compliance Assessment
211	Safety Review or Engineering Change Proposals/Waivers
212	Software Hazard Analysis
213	Government-Furnished Equipment/Government-Furnished Property System Safety Analysis

I do not propose that all the tasks are required in a safety program; however, both the TE manufacturers and equipment users should be fairly knowledgeable about the standard. The safety engineer should understand the language in the event a requirement for a particular task is specified, e.g., if the user is concerned about distinct safety items. MIL-STD-882 provides a detailed explanation of each task. The requirements can be defined in the contract where the supplier will be obligated to respond. Several of the tasks are discussed in this chapter.

One of the principle government agencies that probably affects equipment manufacturers to a great degree is the Occupational Safety & Health Administration (OSHA) and its *OSHA Safety Regulation* report. There are many other federal, state, and local departments, but those deal more with other things such as consumer products, automotive standards and regulations, construction, health, finances, materials, processes, tests, and advertising. There also are many nongovernmental organizations such as the professional societies, independent laboratories, industry associations, and standardization institutions.

Some people believe OSHA standards are much too rigorous, inflexible, and difficult to understand. However, a safety engineer has no choice except to learn all about OSHA requirements, especially for tooling and equipment manufacturers. Everyone wants to work in a safe environment and operate safe machinery. The safety engineer must remain abreast of what is required, and machinery must comply with the regulations.

What are some of the ways in which safety can be introduced into product? Some of these were discussed previously in this book, but there are additional ways to improve safety. A few of these methods are simple; some are complicated. The design engineer and safety engineer will have to weigh the advantages and disadvantages of each condition to determine the extent and justification for the improvement.

Safety Notes in the Manuals

The maintenance manual and operator's manual are used regularly to spread the messages of safety and caution. These comprehensive documents usually are detailed with written instructions, illustrations, and information regarding preventive maintenance procedures, repair and operating instructions, safety notations, and equipment specifications. Safety notations customarily are written in large bold letters, boxed and standing out, and in conspicuous locations. Words such as **Caution**, **Warning**, and **Danger** usually are followed by a description of what demands caution. The safety engineer must be absolutely clear about how the wording should be. There must be no doubt about what a person must be careful.

Sometimes, operators and maintenance personnel believe the manuals are overly simplistic. In some respects, that may be true; however, we should not "throw caution to the wind." When it comes to safety, all personnel must be careful and follow directions that have been thought out, and more often than not, verified and validated. The instructions must be explicit, clear, and detailed to the point they are understood by the intelligence level of the personnel doing the work. Safety engineers must be careful of the wording they use and be certain that they cover all potential safety accidents that could happen. Remember Murphy's Law:

If an accident can happen, it will happen.

Aside from Murphy's Law is the issue of product liability lawsuits, a dread for all manufacturers, especially in our litigious society. Designers and safety engineers must be extremely concerned about potential lawsuits. Again, the caution and warning notes in an operator's manual should be clear and definite about what to do or not do. For example, "Do not stand on the kitchen table to change a light bulb" may seem like a ridiculous statement; unfortunately, such a simple statement sometimes is necessary. Safety engineers must work closely with advertising personnel to ensure that ads do not express actions that should be avoided. Sales exuberance also requires watching because marketing personnel may make promises to customers, without realizing the safety issues in the product. Customers should expect to receive a product that is hazard free, and designers are responsible for ensuring that the product is fail safe. The safety engineer must think about eliminating the causes of injuries at the source. Other departments also have responsibilities in this scenario, which should be supported by top management.

Use of Diagnostics, "Breakpoints," Low-Stress Components, and Protective Arrangements

Chapter 13 discussed diagnostics in considerable length. Monitoring devices were mentioned, but they were not related to safety concerns. Temperature and pressure gages are typical examples of monitoring instruments that can warn operators of impending difficulties and sometimes serious hazards. Tooling and equipment users may not think of these monitors as safety devices, but such monitors do serve as cautionary alarms to ensure that operators at least will be alert to potential problems.

Another method of improving safety is the use of breakpoint or fail-safe parts or devices. Some of these were discussed briefly in Chapter 14. Shear pins, electrical breakers, and fuses are examples of these arrangements. In the fail-safe principle, if a particular system or part is overloaded, a simple replaceable device fails and does not allow a more serious or hazard condition to occur. The electrical fuse or breaker is a perfect illustration of this condition because if the breakpoint part does not fail or if it were not present, a serious fire could take place. At the other extreme condition, a missile abort system, if not present, could play havoc if a misguided firing went astray. Generally, it is a good idea for the design engineer to think about these safety devices for appropriate times when such devices could be utilized.

Similar to breakpoint devices, but not exactly the same, are components that do not require high loads to operate. Chapter 14 discussed derating components related to this type of part. When electronic equipment progressed from vacuum tubes to solid-state apparatuses, it was a godsend to the industry. The components were more reliable, more economical, and better interfaced with other segments of their systems. In addition, they produced less heat, reducing the chances of fire and safety concerns. The idea here is to use components that help lower power levels and reduce heat buildup or other unfavorable conditions.

Planning for ergonomics was discussed in other chapters, but it also is an important criterion for safety. If the man and machine parts relationship is unbalanced, a person could be confused about how to operate a machine, and problems could develop easily. Good shop ergonomics creates safer and more productive work environments, reduces injury claims and absenteeism, and generally enhances morale. When operators do not have to worry about the hazards of the machine, they can focus on quality. The design engineer must concern himself with ensuring that all controls and operating mechanisms are in safe and nonconfusing locations and are well-protected from accidental misuse. Operating and maintenance personnel must know how to utilize the working parts without perplexing conditions or difficulty.

Protective arrangements often are a simple approach to safety. Heat sinks and heat shields are good examples. Extra protection around electrical connectors, wires, or insulators often is overlooked; however, these are relatively simple fixes. Anything that can be done to offset the potential damage from one component to another should be thoroughly evaluated and planned early in the design.

Other Safety Devices

Warranty claims usually do not involve safety, but occasionally they may. Safety concerns should extend well beyond the warranty period. In fact, safety should be a perpetual concern. Manufacturers should be specific about alerting the customer that unsafe practices by the equipment user will not be covered by warranty. Any recalls that are beyond the warranty period and involve safety should be addressed immediately. The government has specific rules in this regard, and the safety engineer should be well aware of those rules.

Training is important with regard to safety. Previous sections of this book discussed the training involved with maintenance and operating personnel. The safety engineer must ensure that documentation, materials, and classroom and on-the-job instructions prevent potential hazards. Safety should be included among all the training R&M requirements. As with other training courses, instructors should be qualified to teach the classes. Experience, knowledge of subject matter, and ability to teach are the criteria that make good instructors.

Summary

This chapter discusses safety concerns of both the user and the manufacturer. The user has a responsibility to specify the reliability of the machinery. This relates closely to safety concerns. This chapter described safety margins and safety factors, which may be slightly confusing. Essentially, the designer should understand the association between the load or stress and the strength of components. The less their distribution patterns overlap, the greater the safety assurance and the reliability will be.

Prioritizing the risk factors and hazards into the four levels of classification helps to organize a system of addressing the factors. The FMEA and FTA were mentioned as two ways in which potential safety problems could be detected before they actually occurred. Chapter 5 presented additional details on these helpful and analytical methods.

This chapter also emphasized the importance of understanding government and industrial standards and regulations and how it is essential for the operator and maintenance manuals to be clear, understandable, and accurate. If operators and maintenance crews do not interpret the manuals correctly, safety can be a moot point. Training puts much of this together. Tooling and equipment manufacturers must remember the determinants mentioned in this chapter. Both users and manufacturers should understand and know what is required.

References

Doty, Leonard A., *Reliability for the Technologies*, 2nd ed., ASQC Quality Press Book by Industrial Press, New York, 1989.

Jones, James V., *Engineering Design Reliability, Maintainability, and Testability*, Tab Professional and Reference Books, Blue Ridge Summit, PA, 1988.

Lewis, E. E., *Introduction to Reliability Engineering*, 2nd ed., John Wiley and Sons, New York, 1994.

MIL-STD-882, System Safety Program Requirements.

Chapter 17

Customer Responsibility

Most of the preceding chapters in this book discussed the designer's and manufacturer's responsibilities for the R&M charges. However, the customer also is accountable. The user must know what R&M parameters are required for his or her machinery and what must be collaborated with the manufacturer to obtain what is specified. In many cases, the customer must work directly with the manufacturer to ensure that the designer of the equipment clearly understands what is required. The *R&M Guideline* matrix of R&M topics matched against the supplier and user presents a fair idea of what is expected from both organizations. User responsibilities and how the user interfaces with the supplier are discussed in this chapter.

Failure Modes and Effects Analysis/Fault Tree Analysis

The user typically does not actually perform an analysis, but the user does provide input to the Failure Modes and Effects Analysis/Fault Tree Analysis (FMEA/FTA) studies. This input may be in the form of requirements or support factors such as failure experience with equipment of similar types, ideas they may have, or items they may have as a concern. These should be discussed with the supplier during and continually after the concept phase. The FMEA/FTA should be initiated as early as possible, preferably in the concept phase.

Quality Function Deployment

The Quality Function Deployment (QFD) reliability improvement initiative, together with all other betterment ideas and methods, can be contributed by the user. Most improvement techniques will generate from the designer, but the user generally also has ideas that should be discussed with the manufacturer. Chapter 14 presented several techniques and methods demonstrating how R&M can be improved. The user should review these to see how he or she can contribute to the enhancement of R&M.

Life Cycle Cost

Life Cycle Cost (LCC) must be a serious concern for any user. The user's chief regards are to have the highest possible reliability at the lowest possible overall cost. The customer must provide accurate input about the environment in which the equipment will operate, the skill levels of operators and maintenance personnel, and the relationship of the cost of the R&M characteristics. The purchase price, consumables, direct and fixed labor, waste-handling, lost production, spare-parts maintenance, life of the equipment, preventive maintenance schedule, cost of repair, breakdowns, and conversion/decommissioning all are costs with which the user is directly concerned. The manufacturer must consider all of these factors when determining the LCC. In some cases, tradeoffs that may be necessary must be negotiated between the supplier and the user.

Qualification Testing

Qualification testing usually is done by the supplier, but such tests also can be accomplished by the user. Much depends on who has the facilities available to do the testing, who believes it can obtain the best information from the testing, who can do the testing most efficiently and effectively, and who can provide the most accurate data from the testing. Prequalification and other early development tests are done more frequently by the supplier rather than the user. After the design is fairly well established and the equipment is manufactured with production tooling (not necessarily production line facilities), the qualification testing can be initiated. The contract will specify agreement about who is responsible for the testing. One of the important necessities during qualification testing is that the test results be forwarded to the manufacturer.

Acceptance Testing

The design of the equipment should be in a mature stage, but it must be validated. In some sense, the manufacturing capability of the design is being tested, although the qualification testing was made on the machinery that was fabricated with production tooling. Nonetheless, problems can occur with the design or manufacturing. This testing should reveal any problems that may exist. The customer usually conducts acceptance testing; however, the testing can be contracted to the manufacturer. It depends on how the user wants the testing to be accomplished. In any event, feedback of information must be sent to the design engineers as in all previous other tests. This must be done to ensure that the supplier is aware of all nonconformances that are present.

Reliability Growth Monitoring

Reliability growth monitoring is primarily a supplier function, but the user has input to the function. The supplier must monitor the R&M progress of the testing to determine the status of the design. The customer certainly should be interested in this progress. Depending on where the phase of development is, the user would become more involved with the growth information as the program matured. This does not imply that the user is less interested in the early development than he or she should be. However, there would be more interest from the user as the design was brought to perfection.

Machine Qualification

If the equipment is in the hands of the user, this is the user's input to machine qualification. The user should know exactly how he or she will use the machine and should take the equipment through all its expected applications. Again, the user may contract the supplier to perform this qualification. However, if that occurs, the customer should be a close observer. In most cases, it would be better for the user to do the actual machine qualification. In any event, the designer must be in the picture because any data generated from the machine qualification must be given as feedback to the manufacturer. The supplier would like to think the design has matured to its completion; however, some difficulties may remain. This information must continue to be furnished to the designer.

Data Collection

The collection of data is primarily a supplier responsibility, but the user should have a significant interest in it. Data collection is the foundation of R&M. Without data collection, there would be nothing to analyze other than conjecture on what is happening to the equipment. The user or person conducting the test program at hand is responsible for generating accurate data on the test performance. If inaccuracies or improprieties exist in the data, it will be notably difficult for the designer to make the necessary corrections to the design. The user or the testing agency has an extremely important responsibility in this regard and should do everything in its power to ensure that the data are flawless, timely, and complete.

Failure Analysis

Failure analysis also is the primary responsibility of the supplier, but the user should have a deep interest in what is happening. Major failures in any program are of great concern, and the customer is probably the most anxious to ensure that the failures are eliminated from the overall design. There is no question that the supplier has the responsibility of performing the failure analysis and then initiating the corrective or preventive actions. The user must have the assurance that the actions have been implemented, tested, and validated, and that the problems in question will not occur the future.

Diagnostic Testing

Training is a supplier function in which the user is the recipient. The customer's responsibility is to assure that his or her personnel understand and become familiar with all the ramifications of the maintenance and repair manual, operating manual, and the diagnostic procedures. The complexities of onboard built-in test equipment and of special external diagnostic service kits should be understood. Diagnostic technicians should know how the equipment can be used not only to diagnose the malfunctions that may occur; they also should know how to report the problem so that the necessary corrections can be made. The supplier provides the training, and the user must assure that its personnel are adept in the diagnostic procedures.

Periodic Maintenance Requirements

As in most other R&M topics, the development of the periodic maintenance requirements is the responsibility of the supplier; however, the user must know what those requirements are and understand how those requirements should be utilized. Most of this information is supplied in the equipment maintenance manual, and training is provided to the user's personnel to ensure that they understand what should be performed. Considerable effort usually is generated in ensuring that the maintenance procedures are effective and essential to the continued operation of the equipment. Training and validation also come into the picture, as there must be assurance that maintenance personnel understand what must be done and how to do it, and that those personnel have the skill levels and knowledge to do what is necessary.

Machine Operation

When the equipment is in the customer's hands, that product is going through its intended operations. If everything was done correctly and all prior testing was completed as it should have been, no problems should occur in machine operation. There may be requirements that the user forgot to specify, or the equipment may be operated differently than the purpose for which it was originally designated. In a sense, this is the final testing, but it should be a kind of wrap-up. All previous tests should have discovered any potential problems. Feedback of machine operation information should continue to be forwarded to the supplier, because improvements nevertheless can be made. However, both the supplier and the user must remember that the further away from concept an R&M program progresses, the more expensive the corrections become. The machine operation is the last phase of the program, and theoretically all problems should have been eliminated by this point. However, if they have not, the manufacturer still must have the opportunity to meet all user requirements.

Training

Both the supplier and the user must be knowlegeable about the R&M requirements and how they can be implemented. The supplier must have all the necessary knowledge and capacity to conduct the R&M program. If the supplier does not possess the skills, it may be necessary to have outside consultants provide training for these proficiencies, or the manufacturer may have to hire personnel who have the qualifications. The training can be extended to both the users and the manufacturers. In some cases, depending on the relationship and rapport between the supplier and the customer, the training can be provided to both parties simultaneously. This can be negotiated between the two organizations.

Summary

Both the supplier and the user have responsibilities in the R&M schedule. The customer basically determines and specifies all requirements that must be met. These requirements must be clearly defined to the supplier. The supplier must understand the requirements and know what is his or her capacity to meet the specifications. The supplier should have the personnel, facilities, and equipment to perform what is necessary to do the job. In turn, the user should be part of the process, starting in the concept phase and continuing to be involved through the disposal phase.

Unfortunately, the user sometimes is not aware of the R&M responsibilities as per the requirements specified in QS-9000TE. When this occurs, the manufacturer should proceduralize its responsibilities in meeting the standard, while at the same time requesting waivers from the customer that release the manufacturer from its accountabilities. This is not an ideal situation, but regrettably this may have to be done in some cases. The topics discussed in this chapter outline the responsibilities of the user, who at minimum should have a knowledge of the concepts that have been presented here. Confidence and trust must exist between the concerned organizations. The customer should know what he or she wants, and this should be documented in the proper contract language to ensure that both parties comprehend what is required.

References

Reliability and Maintainability Guideline for Manufacturing Machinery and Equipment, Society of Automotive Engineers, Warrendale, PA, and National Center for Manufacturing Sciences, Ann Arbor, MI, 1999.

Chapter 18

Reliability and Maintainability Management

Administering the R&M function is a large undertaking in itself, and there has not been much discussion on the subject in this text. There has been a noteworthy exchange about R&M parameters, designer and manufacturer responsibilities, the user's role, and the techniques of R&M. However, what about R&M management? What is the most effective organizational arrangement? To whom should the R&M engineers report? How do R&M personnel relate to quality engineers? How do R&M personnel relate to other departments? In other words, how do the persons with authority create, maintain, and operate the organization to meet their R&M responsibilities and objectives?

We can trace back many years into the history of management, but modern management probably started somewhere in the latter part of the nineteenth century. People such as Henry Towne (1844–1924), Frederick Taylor (1856–1915), Harrington Emerson (1853–1931), Frank (1868–1924) and Lillian Gilbreth (1878–1972), and a few others contributed to the "science of management." They all had their own ideas and endorsements. Some of those ideas still exist; others have become obsolete. Times change, thinking changes, and cultures change.

Many factors are involved in the management of R&M. First, there must be personnel who are knowledgeable about the mathematics and measurements of R&M. They should be engineering personnel or at least have an appreciation and understanding of the engineering disciplines. The life cycle concept must be understood and related to reliability testing and the "cycles" the product witnesses. Documentation must be produced to support the R&M program. Analyses and testing must be conducted. These include the establishment and support for design reviews, life cycle costing, reliability testing, and the feedback of information and its analysis and subsequent corrective and preventive actions.

The company setting up the organizational chart decides how the structure should be with regard to the R&M function and how it relates to the rest of the establishment. Quality and reliability "gurus" have varying ideas on what the relationships should be, and those ideas do not all agree. The following are a few organizational structures that a company can utilize:

- A staff R&M department, be it a one-person operation or multi-person section, reporting directly to the president, general manager, or chief executive officer. The size of the department depends on the size and complexity of the overall organization.

- A line R&M department, reporting to the vice president of operations, who has jurisdiction over several departments, such as engineering, manufacturing, quality, human resources, accounting, and sales.

- A staff or line function reporting to the chief engineer or quality manager.

- A multi-product line organization, with each product having its own sub-departments, including an R&M section.

- A multi-product line organization, except R&M is a staff department, reporting to a position that oversees the "product departments."

The structure of an organization is determined by the owner, chief executive officer, or the president. Much depends on the size of the organization, the skill levels within the organization, the type of work the organization does, customer requirements, the organization's environmental and organizational philosophy, and the organization's goals and objectives. Figure 18.1 provides a pictorial display of the R&M functions and the groups that could be responsible for these charges. However, in any organization, the lines of authority and responsibility should be clear from the top to the bottom. All individuals and departments should understand for what they are accountable. Likewise, they all must know their responsibilities and authority, be as efficient and effective as possible, and receive orders from only one higher authority.

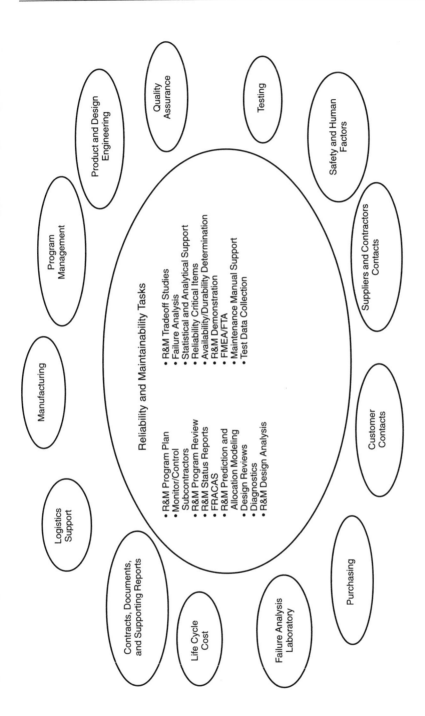

Reliability and Maintainability Tasks

- R&M Program Plan
- Monitor/Control Subcontractors
- R&M Program Review
- R&M Status Reports
- FRACAS
- R&M Prediction and Allocation Modeling
- Design Reviews
- Diagnostics
- R&M Design Analysis

- R&M Tradeoff Studies
- Failure Analysis
- Statistical and Analytical Support
- Reliability Critical Items
- Availability/Durability Determination
- R&M Demonstration
- FMEA/FTA
- Maintenance Manual Support
- Test Data Collection

Product and Design Engineering

Quality Assurance

Testing

Safety and Human Factors

Program Management

Suppliers and Contractors Contacts

Manufacturing

Customer Contacts

Logistics Support

Purchasing

Contracts, Documents, and Supporting Reports

Life Cycle Cost

Failure Analysis Laboratory

Fig. 18.1. Departmental interfaces with reliability and maintenance (R&M) tasks.

Any organizational chart should depict the following:

1. Basic relationship, authority, and communicative lines

2. Who reports to whom

3. Responsible areas of management

4. The type of organizational setup (functional, goal-oriented, matrix, horizontal, or project-oriented)

All types of organizational charts have advantages and disadvantages, depending on the company goals and objectives. Figure 18.1 presented a sample of the types of responsibilities that should be considered in an R&M program. How the organization arranges its operations and controls is its own choice.

However an organization is managed, someone or a department must be responsible for several R&M functions. These include, as a minimum, the following responsibilities:

• Evaluation and allocation

• R&M planning

• Training of R&M personnel

• Proper level of staffing with skilled personnel

• Design and specification reviews

• Monitoring suppliers with respect to R&M activities

• Auditing R&M activities

• Providing service and consulting to other departments, customers, and subcontractors

- Performing tradeoff studies

- Collecting and analyzing data

- Ensuring initiation of corrective and preventive action

- Overseeing R&M demonstration and growth monitoring

- Participating in project and problem-solving teams

- Submitting R&M reports to management and other concerned departments

As a part of the organization structure, certain skills are necessary in an R&M section. We can argue about "degreed engineers" versus "non-degreed" personnel where the president of the organization can decide what is acceptable. In any event, R&M engineers must either have or be able to oversee the following main aptitudes and abilities:

- Perform an analysis of customer requirements.

- Possess and understand the skills of statistical analysis; Weibull analysis; Statistical Process Control (SPC); Failure Reporting, Analysis, and Corrective Action System (FRACAS); problem-solving tools; basic R&M; and the other disciplines discussed in previous chapters.

- Develop and evaluate R&M prediction plans.

- Develop and manage R&M growth monitoring procedures.

- Understand the environmental impact on components, subsystems, and systems.

- Receive, provide, and analyze information on part failures.

- Know how to evaluate the R&M parameters of a product.

- Develop and utilize R&M evaluation models and methods.

- Develop R&M demonstration and evaluation procedures.

- Monitor vendor R&M programs.

- Participate in design reviews.

- Provide consulting service to others.

Note that many of the individual characteristics of R&M engineers are similar to the responsible functioning of an R&M section. The R&M department should contain individuals who have the skills necessary to fulfill the department function. In addition to these attributes, a group or individual must oversee the maintenance program. This can be accomplished within the R&M group or outside the section. One of the main functions of the maintainability group is to write the maintenance manual(s), which in itself usually is an enormous task. The manual procedures must be verified and validated, and operators and maintenance personnel must be trained. Support must be provided for the R&M tasks and maintainability personnel. All this and a persistent updating and correcting of the manual must take place.

Other areas of R&M management extend into departments beside the R&M section. This includes project and design engineers, quality engineers and technicians, human factors and product safety personnel, and other department managers. There should be no question about upper or executive management support. Top-level buttressing and encouragement must be apparent and practiced. If executive management does not support the R&M program, the program can fall into disarray. Figure 18.1 portrayed R&M activities connecting with various other departments within an organization. Each company must decide on the complexity of its own organizational interfacing, but the chart depicts how R&M can relate to other departments.

Project and design engineers must be alert to the performance of their designs at various times. This is especially true when failures occur. Designers must be able to accept that their designs had inadequacies. The purpose of testing programs is to determine the reliability of a specific product and any potential problems. These problems usually appear during testing, after which the design or project engineer must determine why the product failed. Failure analysis must be performed to determine if the problem was related to design, manufacturing, or a supplier part. The following questions should be asked, in one form or another:

- Was the product to specification?
- Was the test within the test parameters?
- Did operating personnel follow written procedures?
- What were the environmental conditions?
- How did the product actually fail?
- What were the circumstances behind the failure?
- What would improve the reliability?

When failure analysis reveals an unquestionable cause of the failure, responsibility for corrective or preventive action usually is straightforward. However, if the cause is not clear, "finger pointing" often occurs regarding who is responsible for the fix. It is not easy for a design engineer to admit that his or her design did not perform to customer requirements. This is especially true if the designer spent a considerable amount of time in developing the design. To determine an answer, the reliability engineer frequently must spearhead the analysis effort through team involvement and cooperation. Likewise, management must support this effort.

Engineers also should keep in mind the Analyze-Test-Analyze-And-Fix (ATAAF) principle, rather than the Test-Analyze-And-Fix (TAAF) principle. Whatever designers can do to predict failure modes and then design those potential failures out of their machinery is a definite plus. Sometimes, designers think their designs are acceptable and believe the testing will verify what they have produced. This could be true; however, it would be much better if the designers could prevent failures in the first place. This also means that the designers should not overdesign or go beyond the limits of life cycle costing. The FMEA/FTA and reliability prediction techniques can prove extremely useful in determining possible failure modes before they occur.

Other disciplines also become involved with the R&M effort. Failure analysis usually is conducted in an analytical laboratory, where a skilled failure analyst usually can pinpoint the cause of a failure. This information is essential to the

R&M engineer and design engineer because it provides a clue about what corrective action should be taken. An organization either must have personnel who can provide this service or must be able to obtain the service from outside laboratories.

If a problem is traced to a manufacturing anomaly, the quality and manufacturing engineers frequently are involved. They must determine the reason why the part was produced incorrectly and then initiate the corrective and preventive action to ensure that the problem does not recur. Unfortunately, manufacturing and design engineers (and quality personnel) sometimes do not agree on why a particular part failed. The root cause of a failure must be determined in one way or another. A failure analysis team, a Failure Review Board (FRB), or any other qualified group generally is the best approach to determining the reason for the failure.

Test and maintenance personnel also are key members of the R&M effort. These workers often are the only ones who can provide details about the events that either caused or led to the failure. Their reports to the R&M department must be accurate and contain enough detail so that the information is useful in the failure analysis. Often, the reliability engineer or designer may have to discuss the problem with the test or maintenance personnel. The idea is to obtain as much information as possible concerning the failure and its possible cause. Then, using this information and other accumulated data, a judgment can be rendered concerning the fix.

The life cycle manager or someone with similar responsibilities must follow the development process from concept to disposal. He or she must work with the R&M engineer and other personnel to determine what costs are involved. Cost of operations, testing, anticipated failure analyses, inspections, manual generation, time and personnel involvement, disposal, safety requirements, facilities, documentation necessities, and parts control are some of the tasks of which life cycle personnel must be aware. The R&M engineers must work closely with life cycle personnel to keep costs at required, competitive, and reasonable levels.

Other departments involved in the R&M process include document control, human resources, safety, logistics, value engineering, contracts, finance, quality, manufacturing engineering, and last, but certainly very important, management. All have a significant input to the R&M effort. Also part of the management endeavor are activities that include identifying the cost drivers, establishing and estimating the costing at various phases, determining how costs can be reduced, and assessing costs for individual components or assemblies.

In summary, R&M management has no one special or efficient organizational setup because much depends on the size of the organization, the complexity of the product, the skill and knowledge of personnel, and the level of support from management. If an organization does not receive such crucial support from management, the R&M program will not be fully effective. "Lip service" is not enough. There must be *bona fide* and committed support with sincere effort in meeting or exceeding customer requirements.

As we review all the chapters of this book, we should be able to recognize the advantages and benefits of an R&M program. Even if a supplier is not contractually required to have such an arrangement, a good R&M program offers remuneration to both users and manufacturers. An R&M program provides a disciplined approach to defining requirements and listening to the voice of the customer. Analytical studies provide a savings of labor and paperwork, as well as verification of the requirements. It is reasonable to expect that the end product will be superior in function and will have been produced at the lowest possible cost. More effective communication and interfacing between engineering design and the other functions within the organization also will occur. From an overall viewpoint, the utilization of a good R&M program should be a "win-win situation" for all parties involved.

References

Dhillon, Balvir S., and Reiche, Hans, *Reliability and Maintainability Management*, Van Nostrand Reinhold Co., New York, 1985.

Doty, Leonard A., *Reliability for the Technologies*, 2nd ed., ASQC Quality Press Book by Industrial Press, New York, 1989.

Jones, James V., *Engineering Design Reliability, Maintainability, and Testability*, Tab Professional and Reference Books, Blue Ridge Summit, PA, 1988.

List of Acronyms

AL	Applied Load
ANSI	American National Standards Institute
ASQ	American Society for Quality (formerly ASQC, American Society for Quality Control)
ATAAF	Analyze-Test-Analyze-And-Fix
Big Three	Chrysler Corporation (now Daimler-Chrysler AG), Ford Motor Company, and General Motors Corporation
BIT	Built-In Test
BITE	Built-In Test Equipment
C/I	Cause/Influence
ES	Engineering Specification
ESS	Environmental Stress Screening
F&T QSS	Ford's Facilities and Tools Quality Systems Survey
FA	Failure Analysis
FEA	Finite Element Analysis
FMEA	Failure Mode Effects Analysis; *also* Failure Modes and Effects Analysis (same as FMECA)
FMECA	Failure Modes Effects and Criticality Analysis (same as FMEA)
FRACAS	Failure Reporting, Analysis, and Corrective Action System
FRB	Failure Review Board
FTA	Fault Tree Analysis
GIDEP	Government Industry Data Exchange Program
ISO	International Organization for Standardization
ITI	The Industrial Technical Institute

KISS	Keep It Simple, Stupid
LCC	Life Cycle Cost
MMBF	Mean Miles Between Failure
MMBPM	Mean Miles Between Preventive Maintenance Actions
MR	Maintenance Ratio
MTBF	Mean Time Between Failures
MTBUM	Mean Time Between Unscheduled Maintenance
MTPPM	Mean Time to Perform Preventive Maintenance
MTTF	Mean Time To Failure
MTTR	Mean Time To Repair; *also* Mean Time to Replace
NCMS	National Center for Manufacturing Sciences
NGT	Nominal Grouping Technique
NIST	National Institutes of Standards and Technology
NR	Component Net Rating
OSHA	Occupational Safety & Health Administration
PDCA	Plan, Do, Check, Act Cycle
PDPC	Process Decision Program Chart
PDSA	Plan, Do, Study, Act Cycle
PERT	Program Evaluation and Review Technique
PM	Preventive Maintenance
PPAP	Production Part Approval Process
QFD	Quality Function Deployment
R&D	Research and Development
R&M	Reliability and Maintainability
RADC	Rome Air Development Center

RBD	Reliability Block Diagram
SAE	Society of Automotive Engineers
SF	Safety Factor
SPC	Statistical Process Control
TAAF	Test-Analyze-And-Fix
TE	Tooling and Equipment
TESQA	Tooling and Equipment Supplier Quality Assurance

Index

Mean Time Between Failures (MTBF)
  in reliability prediction, 121, 122
  from typical Weibull distribution, 86-87, 94, 95
  utilizing Reliability Block Diagram (RBD), 124
  See also R&M prediction/allocation
Mean Time To Repair (MTTR)
  defined, 40
  as measure of system repairability, 23
  QS-9000TE requirement for, 44, 83
  specified in *R&M Guideline,* 44
  subsystem repair times, addition of, 122
  See also R&M prediction/allocation
Memory Jogger II, 187, 189
MIL-STD-785B
  ESS requirements in, 81
MIL-STD-882 (System Safety Program Requirements)
  100 series/200 series tasks in, 198-199
Mitsubishi Kobe Shipyard
  and Quality Function Deployment (QFD), 67
Monte Carlo modeling, 126
MTBF *See* Mean Time Between Failures (MTBF)
MTTR *See* Mean Time To Repair (MTTR)
Multiplication Law, 24
Murphy's Law, 200

National Center for Manufacturing Sciences (NCMS)
  develops R&M guidance document, 2
  publishes *Reliability and Maintainability Guideline...,* 1
Nominal Grouping Technique (NGT)
  as problem-solving technique, 175-176
Normal distribution, curves for, 25

OSHA standards (and product safety), 199

Pareto chart, 169-170
Plan, Do, Check, Act (PDCA) cycle, 177, 179-180
Poisson distribution, curves for, 25
Poka-Yoke (idiot proofing/mistake proofing), 154

About the Author

George Mouradian's background spans almost 50 years of experience in quality and reliability in various industries. He has held quality and reliability engineering positions with AM General Corporation, Rockwell International, Consumer Power Company, Massey Ferguson, LTV Aerospace, Aerojet General Corporation, Wolverine Tube Corporation, Ford Motor Company, the U.S. Army Ordnance Corporation, and Cadillac Motor Car Division.

Mr. Mouradian currently is a quality/ reliability consultant specializing in QS-9000TE preassessments and registration and in assisting organizations in their quality and reliability programs. He has guided many companies in achieving their QS-9000 registrations and their ISO registration goals. Mr. Mouradian has directed and managed quality and reliability programs; determined and predicted reliabilities; established and maintained a Failure Reporting, Analysis, and Corrective Action System (FRACAS); performed Failure Modes, Effects, Criticality Analysis (FMECA); conducted design reviews, reliability testing, and failure review boards; and wrote reliability program plans and QS-9000 procedures. In 1997, he also taught reliability engineering and quality management at the American University of Armenia in Erevan, Armenia.

Mr. Mouradian has B.S. degrees in both mathematics and industrial engineering and an M.E. degree in statistics, all from Wayne State University. He is a Professional Quality Engineer in the State of California, a Registration Accreditation Board Lead Auditor, and an International Auditor and Training Certification Association Senior Auditor. Mr. Mouradian is an active member of the American Society for Quality, the Society of Reliability Engineers, and the American Society of Materials, and he is past president of the Armenian Engineers and Scientists of America. He has published numerous articles and books, including the *Certified Quality Refresher Course Manual*, *ISO/QS-9000 Internal Auditor Manual*, and *Armenian InfoText* (a mini-encyclopedia about the Armenians).

241